Theodore T Groom

On the early development of Cirripedia

Theodore T Groom

On the early development of Cirripedia

ISBN/EAN: 9783741140143

Manufactured in Europe, USA, Canada, Australia, Japa

Cover: Foto ©Klaus-Uwe Gerhardt /pixelio.de

Manufactured and distributed by brebook publishing software
(www.brebook.com)

Theodore T Groom

On the early development of Cirripedia

V. *On the Early Development of Cirripedia.*

By Theodore T. Groom, *B.A., B.Sc., F.G.S., late Scholar of St. John's College, Cambridge.*

Communicated by Adam Sedgwick, *F.R.S.*

Received May 11,—Read June 16, 1892.

[Plates 14-28.]

Table of Contents.

Part I.

Part II.—Embryonic Development.

                                            12.3.94

PART III.—THE FREE NAUPLIUS, FIRST TWO LARVAL STAGES.

PART VI.—BIBLIOGRAPHY.

PART I.

INTRODUCTION.

During a short stay at Plymouth, in 1889, I was engaged in studying certain points in the anatomy of Cirripedia; finding, however, that a knowledge of the embryology was necessary in order to arrive at a complete understanding of the adult structure, I became wishful to investigate the life-history of some one member of the group. This I had an opportunity of doing at Naples, where I was appointed to occupy the Cambridge University Table at the Zoological Station for a period of six months, subsequently increased to nine. I here succeeded in obtaining a practically complete series of stages of *Balanus perforatus*, BRUGUIÈRE, as well as many stages in other members of the group.

Though a number of able observers have occupied themselves with the embryology of Cirripedes, yet, owing to lack of opportunity, and to the difficulty of obtaining complete series of developmental stages, as well as to the inherent difficulties in the subject, much remained to be done in this line. WILLEMOES-SUHM alone, with the advantages afforded by his position during the Challenger Expedition, has hitherto obtained a complete series of stages of any one form, but he failed to trace the history of the earlier stages, and in the later, limited himself to the appearance of fresh and spirit specimens, as seen without cutting sections. In fact the method of sections has been little applied to the development of Cirripedes, and not at all to the earlier stages. There is, therefore, little apology needed for an account embracing the results obtained by the employment of some of the more modern methods of embryological study.

My account of the development of *Balanus perforatus* differed in so many respects from that of previous observers on the same and other Cirripedia, that I was induced to compare the development in a number of forms, believing that the wide differences stated to exist must be due, at any rate in part, to actual differences in the development of the different forms, such as occurs in other groups in which the ontogeny of allied species and genera have been compared. This expectation was, however, only partially realised; I found, on the contrary, a great uniformity in the development of the different forms, an agreement, in fact, so close that all might be conveniently treated together. The species studied were *Balanus perforatus*, *Chthamalus stellatus*, POLI; *Lepas anatifera*, LINN.; *L. pectinata*, SPENGLER; and *Conchoderma virgata*, SPENGLER. I was enabled by a comparison of these forms to confirm previous work on certain points, and to establish other facts which have, in many cases, the merit of reconciling apparently discordant opinions. The present account treats of the development of these forms as far as and including the second Nauplius stage.

The early development of the Rhizocephalan genus *Peltogaster* was also investigated, but as this form showed certain differences from the others, I judged it best to reserve my account of it for a future occasion.

MDCCCXCIV.—B. R

(A.) METHODS OF OBTAINING THE OVA AND EMBRYOS, AND OF EXAMINING THE
OVA, EMBRYOS, AND NAUPLII.

The ova are readily obtained by cutting open the shell of the adult with a strong
pair of scissors; the ovaries may then be distinguished through the membrane of the
mantle cavity, or the ovigerous lamellæ may be seen lying at the bottom and sides of
the cavity, free in Balanids, but attached to the *ovigerous frena* in Lepads. The
lamellæ may be taken out by a pipette in the case of the smaller species, or preferably
by a small section-lifter or pair of forceps; the latter are necessary in the case of
the Lepads.

Placed in watch glasses the development may be watched from time to time. I
found it impossible, however, to trace the whole history of the development of the
Nauplii in this way, the process invariably ceasing after a certain time for want of
some condition I failed to ascertain, though I tried many methods of culture. . Species
perhaps vary in this respect, since MÜNTER and BUCHHOLZ (22) succeeded in watching
the whole course of embryonic development in one lamella of *Balanus improvisus*, DARW.
In consequence of this circumstance I was compelled to examine the ova of a number
of animals found at different stages, in order to complete the account. This was
rendered more difficult by the fact that, as is well known, the ova of the two lamellæ
of any animal are nearly always at practically the same stage of development: one
can consequently search for certain stages for a long time without success. Thus,
though I examined the eggs of thousands of individuals of *Balanus perforatus*, I never
succeeded in obtaining an ovum, the formation of the blastoderm of which had not
commenced, having thus the same experience as LANG, whose account of segmentation
of this form commences when the first blastomere (his ectoderm) is already separated
from the yolk (his endoderm). The mechanical labour of cutting through a solid
shell, such as that of *Balanus*, becomes in such cases very tedious and occupies much
time, and my best thanks are due to Dr. EISIG for supplying me with assistance in
this matter. In other forms a number of stages were sometimes found in the same
lamella, as in *Conchoderma virgata*, which agrees more with *Balanus improvisus* in
this respect.

The methods of obtaining the Nauplii are given later.

The various stages were first of all examined by transmitted and reflected light.
The embryos are sufficiently transparent, especially in some species, to show most of
the details of their anatomy by transmitted light. ABBE'S condenser was frequently
found to be of considerable assistance for this purpose, the oblique light often being
very necessary, in order to make out the cell boundaries. It was, however, always
necessary to rotate the egg by moving the cover-glass along, as eggs were rarely
sufficiently transparent to show the details on both sides. Some eggs, such as those
of *Lepas* and *Conchoderma*, were easily detached from one another and rotated; others,
such as those of *Balanus perforatus*, were often exceedingly obstinate, in consequence

of the firm character of the cementing material between the eggs. For the later embryonic stages reflected light was essential, and a condenser was found of great use in determining the boundaries of the embryonic organs.

For the histology of the embryos picro-nitric, picro-acetic, and picro-sulphuric acids and PERENYI's fluid were good. PERENYI's fluid was found to be a generally useful reagent, and preserved all stages well, but took longer to stain than the rest. Picro-acetic acid was also good for the embryonic stages, and, followed with borax-carmine, gave excellent results. Picro-sulphuric and picro-nitric specimens, stained with DELAFIELD's hæmatoxylin, were also excellent, the former for the embryonic, and the latter for the free Nauplius stages.

For examination of unstained Nauplii weak osmic acid, as recommended by HOEK, was useful, as was also weak iodine, especially in the case of *Lepas* and *Conchoderma;* the former could be immediately followed by BEALE's carmine, in the way recommended by GROBBEN (35), but some maceration inevitably followed. In addition to these, the best reagents for preserving the form were corrosive sublimate, alone or with acetic acid, PERENYI's fluid, and chromic acid ; embryos fixed by the latter reagent were, as might be expected, difficult to stain; but Nauplii from the two former fluids stained well with borax-carmine for a short time (a few minutes to a quarter of an hour).

Examination of preserved examples of the embryonic stages gave few results in consequence of the presence of the vitelline membrane, and its resistance to staining reagents.

For sections it was simply necessary to take a small piece of an ovigerous lamella preserved according to one of the above methods, imbed in paraffin, and afterwards stain on the slide. In the case of borax-carmine the staining was best done before imbedding. A large number of serial sections of all early stages of *Lepas* and *Balanus* were obtained by the aid of the Cambridge rocking microtome.

(B.) Seasons at which the Ova, Embryos, and Nauplii are found.

Before giving an account of the mode in which the eggs are laid, it may be well to state where the species may be conveniently found, and the seasons of the year at which development takes place.

The limestone rocks beneath the Marine Biological Laboratory at Plymouth, are covered in places, even considerably above high-tide mark, by small Balanids ; these consist chiefly of *Chthamalus stellatus*, the eggs of which may be found in countless numbers during the months of July and August, and probably at other times.* At

---

* Nauplii of this species, together with those of a species of *Balanus*, were frequently obtained at large numbers at Plymouth by the use of the tow-net, during the months of February and March in 1893 ; by the beginning of May they had become distinctly rare ; they also appear to be rare or absent at other times of the year. A few Cypris-stages, probably belonging to this form, were obtained from tubes sent from Plymouth in the middle of May, 1892.—[10/7/93.]

Newquay, in Cornwall, I found eggs also in September. At Naples the same species is common everywhere a little above the sea-level ; it probably breeds here the whole year round.

Near low-tide level, at Plymouth and Naples is the larger *Balanus perforatus*, breeding also probably the whole year round at Naples, all stages being found at any time examined. Of its breeding at Plymouth I have only a single note ; eggs were found at the beginning of June, the only time I looked for them.

These two species are convenient for studying the embryonic stages, the former from the ease with which all stages can be obtained, and the latter from the clear separation of blastoderm and yolk, and the distinctness with which the nuclei of the yolk-cells can be seen without special treatment.

In the case of *Balanus balanoides*, which is sparsely scattered about at Plymouth and Newquay among the *Chthamalus* (and sometimes hard to distinguish from it), but very abundant at Worm's Head, in Glamorgan, I found no eggs in July, August, or September, these observations agreeing with HOEK's statement (30) that the development lasts from November to February, or later.* I may add that I could not find this species in the Bay of Naples, a fact agreeing with DARWIN's belief that it does not extend into the Mediterranean.

For ova of *Lepas anatifera, L. pectinata,* and *Conchoderma virgata,* one has to rely on floating pieces of timber, ship's bottoms, &c. All stages of development were found at Naples, at any time during the year the animals were brought in.

It appears, thus, that the Cirripedes do not behave uniformly with respect to the time and duration of development. While some, such as *Balanus perforatus,* breed all the year round, others, such as *Balanus balanoides,* have one period of development, at any rate in certain localities. According to HESSE (12), *Scalpellum obliquum* breeds in the summer ; I failed also, at Naples, between the months of October and May, to find eggs in *Balanus amphitrite,* though many examples were brought me, so that the breeding of this species also possibly takes place only in summer. DARWIN's supposition that Cirripedes breed several times a year (10) must probably be replaced by the statement that Cirripedes breed throughout the year, or during only a portion of it.

Some observations may be here added which it was not possible to incorporate in the table. Nauplii of *Balanus balanoides, B. porcatus, Chthamalus,* and *Verruca Strömia,* as well as Cypris-stages of *Chthamalus stellatus,* were obtained by SPENCE BATE (8), in Devonshire, during the summer.

The development of *Pollicipes polymerus* was partly worked out by NUSSBAUM, in California, during the dry season.

* Nauplii and Cypris-stages of this species appear to be abundant at Jersey, at times, during the spring, but at no other part of the year. I received examples taken by Messrs. SINEL and HORNELL in March, 1893 ; by the beginning of May they had again practically disappeared.—[13/7/93.]

The following table shows the months in which eggs, Nauplii, or Cypris-stages, have been obtained by myself or others. The capitals refer to the places of observation, and the numbers to the observers.

	January.	February.	March.	April.	May.	June.	July.	August.	September.	October.	November.	December.
Lepas anatifera	: N⁹	N³·⁵	N⁷	N⁷·⁹	N⁸·⁹	: :	: :	: : : :	Q⁹	N⁹·⁹	:	N⁹
" pectinata	N⁹	: :	N⁷P⁸·¹	: :	N⁷·⁹	: P²·⁴	: P²·⁴	: : : :	Q⁹	N⁹	:	
" fascicularis	N⁹	N⁹	: :	: :	: N⁷	B¹ : :	B¹	B¹	:	N⁹	N⁹	N⁹
Conchoderma virgata	: :	: :	: :	: :	B¹			P¹	B¹ :	B¹		B¹
Scalpellum vulgare	: :	: :	N⁹P⁹	: N⁹	N⁷P⁹P⁹		B¹	P⁹	B¹ Q⁹			
" obliquum	N⁹·⁹·⁹	N⁹P⁹	N⁹P⁹	N⁹·⁹	N⁹P⁹	B¹ : : P⁸	P⁹P⁹P⁹Q⁹	: P⁸	: P⁸	N⁹·⁹	N⁹	N⁹·⁹
Chthamalus stellatus	N⁹·⁹·⁹	N⁹·⁹	N⁹·⁹	N⁹·⁹	N⁹·⁹	: P² : :	: : :	: P²	: P²	N⁹·⁹	N⁹·⁹	N⁹·⁹
Chenolobia patula									R⁹	R⁹	R⁹	R⁹
Balanus perforatus	L⁵	: L⁵	: :	: :	: :	: : :	o (W⁹)	o (Q⁹)	o (Q⁹)	:	L⁵	L⁵
" improvisus							o (Pⁱⁱ)	o (Pⁱⁱ)				
" balanoides												

Place of Observation.

B. Brest. (?)
F. Frontignan.
L. Leiden.
N. Naples.
P. Pacific Ocean, 35° N.
Pl. Plymouth.
Q. Newquay (Cornwall).
R. Mouth of R. Ryck (Pomerania).
W. Worm's Head, Glamorganshire.

* Cypris-stage observed.
o. Ova and embryos not found, though looked for at the places indicated in brackets.

Observer.

1. Hesse.
2. Pagenstecher.
3. Münter and Buchholz.
4. Willemoes-Suhm.
5. Hoek.
6. Lang.
7. Schmidtlein.
8. Lo Bianco.
9. Groom.

(C.) OVIPOSITION, FERTILIZATION, AND FORMATION OF THE OVIGEROUS LAMELLÆ.

It is well known that the ova of Cirripedes are to be found at the sides of the body in the mantle cavity, cemented together into more or less extensive lenticular sheets, termed by DARWIN the "ovigerous lamellæ."

The mode in which these are formed has long been the subject of discussion, no one having observed the act of oviposition.

The ovaries were recognised as early as 1835 by WAGNER (52), MERTENS (53), and MARTIN SAINT ANGE (54). The latter supposed the ova to pass into the mantle cavity by means of an aperture under the carina. DARWIN believed, on the other hand, that the "true ovaria" were represented by glandular organs, described by MARTIN SAINT ANGE as appendages to the stomach, and lately shown by HOEK to be really such. The oviducts, according to DARWIN (9, 10), passed from the true ovaria ("enteric diverticula") in the capitulum past the base of the first pair of cirri into the peduncle, to end in the branching ovarian tubes (ovaries proper); from these the ova were believed to burst into the corium (immediately under the cuticle lining of the mantle cavity), which "resolves itself into the very delicate membrane separately enveloping each ovum, and uniting them together into two lamellæ." Upon exuviation the ova passed into the mantle cavity, while a new cuticle formed beneath.

KROHN (53), recognising with the earlier observers the ovaries proper, traced the oviduct in the reverse direction to the base of the first pair of cirri, where they opened. The ova, he supposed, made their exit through this opening.

PAGENSTECHER (15) did not believe with KROHN that the oviducts could be traced to the first pair of cirri, but rather that after passing up the peduncle they opened beneath the labrum.

KOSSMANN (56), HOEK (39), and NUSSBAUM (51), on the other hand, have confirmed the anatomical description of KROHN, and NUSSBAUM (44) has shown that the ova are at first placed at the mouth of the oviduct, becoming afterwards detached. NUSSBAUM also detected ova in the oviduct itself.

I was able at the beginning of the past year to complete the evidence on this question by finding an individual of *Lepas anatifera* in the act of oviposition. The ovigerous lamella of one side had been completed, and the eggs examined from time to time were seen to be developing in the usual manner; but, on the other side, the eggs had only just commenced to issue from the mouth of the oviduct, and formed a small soft rounded gelatinous mass continuous with a string of ova passing into the mouth of the oviduct, and filling the sac at the base of the first pair of cirri. The eggs of *Lepas anatifera* being of a beautiful blue colour, it was quite easy to trace the course of the oviduct, now distended with eggs, all the way from the sac at the base of the cirri through the prosoma to the ovary in the peduncle; where, not visible at first, slight pressure revealed its existence, and its course was seen to be precisely that indicated by the last-mentioned observers.

It may therefore be regarded as firmly established that the ova arising, as described by KROHN, KOSSMANN, MÜNTER and BUCHHOLZ, HOEK, and others, in the branched ovary, pass along the oviduct to the base of the first pair of cirri, and are ejected thence into the mantle cavity, where, cemented together, they form the ovigerous lamella.

As to the method of fertilization, DARWIN (10), FRITZ MÜLLER (55), and SPENCE BATE (21) have given facts showing that cross fertilization must take place in some cases. When cross fertilization fails the position of the penis seems eminently suited for self-fertilization, as inferred by MARTIN SAINT ANGE (54).

As to exactly at what moment fertilization takes place I have little fresh evidence. NUSSBAUM (44) supposed at first that it occurs in *Lepas* before the formation of the egg-sacs, and, therefore, while the eggs are still in the oviduct; later, however (51), he supposed that the ova were fertilized as they issued from the external opening of the oviduct; the fact that in the foregoing case of *Lepas anatifera* the ova of the completely developed lamella went on developing quite normally, while those in the small clump emitted on the other side did not develop further, though united together by the gelatinous cement and placed in exactly similar conditions, tends also to show that fertilization takes place in the mantle cavity. In all other cases (and these were fairly numerous) where I obtained eggs of Cirripedes freshly laid they had already commenced developing, and had evidently been fertilized. I am inclined to believe, therefore, that the spermatozoa do not enter the oviducts, and that fertilization takes place immediately after extrusion of the eggs, while the cementing material is still quite soft and easily penetrable.[*]

The ovigerous lamellæ vary much in size, colour, and number, not only in the different species, but in different individuals of the same species; they also, as has been previously observed, in certain cases undergo definite changes in one and the same individual.

With regard to the first point, the size of the lamella is, roughly speaking, proportional to that of the individual. In all the species much variation in size and shape may be noted; the lamella may at times be very small and tolerably thin, or may form large sheets filling up a considerable portion of the space between the body and the walls of the mantle cavity.

The newly-formed lamellæ of *Lepas* and *Conchoderma* are of a beautiful deep blue colour, and form a rounded gelatinous mass at the opening of the oviduct. As development proceeds the mass expands in all directions, and becomes a more or less extensive sheet of ova firmly cemented together; at the same time the colour changes successively from blue into purple, red, and pink. From the colour a very good idea can be obtained of the stage of development of the ova, and though a particular tint

---

[*] NUSSBAUM describes the material which cements the eggs together in *Pollicipes* as perforated, and suggests that the perforations serve as passages for the spermatozoa.

by no means always indicates precisely the same stage of development, this fact is exceedingly useful.

In *Scalpellum obliquum* (12), and apparently in *Pollicipes polymerus* (49), the lamellæ are yellow. In *Dichelaspis Darwinii* the ova are vermilion-red (17). In *Chthamalus stellatus*, the ova, at first flesh coloured, become bright orange-red, and the lamellæ pass gradually from this colour into paler orange and chrome-yellow. In *Balanus perforatus*, the lamellæ, at first pale yellow, become successively yellowish-grey and grey. BOVALLIUS (26) has noted a similar gradation of colour in the *Balani* he studied.

In the table, on page 130, the sizes of the ova examined are given, together with a few measurements previously published. My own measurements refer to the peri-vitelline membrane which is formed at the same corresponding period in all the species ; this method of measurement was found to be the most convenient one owing to the difficulty of determining fixed points in the actual embryo. The membrane does not increase much, if at all, in size till the Nauplii are approaching maturity ; any increase in growth is masked by the variations in size which occur. The vitelline membrane of the same species was repeatedly measured at different periods ; the different batches (the ova of each batch being mostly at the same stage of development) often varied in average size, but quite independently of the stage reached.

From the table it is seen that the ova vary in average length from 0·11 millim. in *Balanus improvisus* to 0·31 millim. in *Balanus balanoides*, and in breadth from 0·08 millim. in *Dichelaspis* to 0·19 millim. in *Balanus balanoides*. In the species I observed myself the range was considerably less ; the eggs varied in length from 0·132 millim. in *Lepas anatifera* to 0·202 millim. in *Balanus perforatus*, and in breadth from 0·084 millim. in *Chthamalus* to 0·132 millim. in *Conchoderma*. The average length in species other than those of *Balanus* ranges between 0·164 millim. and 0·17 millim., and the breadth between 0·08 millim. and 0·12 millim.*

At first, as HOEK found in *Balanus perforatus*, no increase in size in the egg takes place, but as they approach maturity the increase in size of the embryo causes elongation of the vitelline membrane in all the species I examined. In cases where the Nauplii were nearly ready to hatch, the size of the ovum was 0·25 millim. by 0·16 millim. in *Lepas anatifera*; 0·19 millim. by 0·095 millim. in *Chthamalus stellatus*, and 0·24 millim. to 0·27 millim. in length by 0·127 millim. to 0·139 millim. in breadth in *Balanus perforatus;* and 0·16 millim. to 0·17 millim. in *Balanus improvisus* (22); an increase amounting in most cases to one-half of the original length.

* DARWIN gives the size of the ova in some genera, those of *Chthamalus* and *Scalpellum* measuring respectively 0·176 millim. to 0·578 millim. In *Lepas* the length varies from 0·176 millim. to 0·226 millim., and in *Balanus* from 0·239 millim. to 0·314 millim. But it seems probable that the latest stages are included in these measurements, and that the measurements in consequence cannot be precisely compared with those in the table. They are, however, practically within the ranges given in the tables for the corresponding genera.

Apart from this increase of size, the egg of *Lepas pectinata* is thus a little smaller than that of *Lepas anatifera*, especially in breadth, and is rather more ovate. In *Conchoderma virgata* the eggs are a little larger than those of *Lepas anatifera*, especially in breadth. In *Dichelaspis Darwinii*, according to FILIPPI, the eggs are smaller than in the foregoing genera, especially in breadth, a fact agreeing with the smaller size of the Nauplius when first hatched. The eggs of *Chthamalus* are longer than those of the foregoing genera, but as in *Dichelaspis*, narrower than in *Lepas* and *Conchoderma*. In *Balanus perforatus*, as above stated, the eggs are generally longer than in any of the foregoing, but of a breadth intermediate between that of *Lepas anatifera* and *Lepas pectinata*. The egg of *Balanus improvisus* is apparently shorter than those of the foregoing, but of the same width as *Chthamalus*; while that of *Balanus balanoides* is larger in both directions than any of the rest. In this genus, as in the others, and generally in the group, the size of the egg has a distinct relation to that of the Nauplius, but none to that of the adult.

In the table is also given the range of variation in the size of the ova. The variation is greatest in *Balanus perforatus* and *Lepas anatifera*, and least in *Lepas pectinata*.

The ova of each ovigerous lamella show a certain amount of variation, the range of which varies with the species and individual ; thus five ova of *Balanus perforatus* at an early stage taken from one lamella varied in length between 0·176 millim. and 0·208 millim., while five older eggs taken from the lamella of a different individual varied in length 0·195 millim. and 0·202 millim.; in breadth the ova of the former batch varied from 0·113 millim. to 0·12 millim., while those of the latter varied from 0·094 millim. to 0·107 millim. The average length of both the batches is 0·195 millim., while the average breadths are 0·115 millim. and 0·101 millim. Longer ova are sometimes correspondingly less in width ; thus, an egg 0·176 millim. in length was 0·12 millim. in breadth, while one 0·195 millim. in length, measured 0·107 millim. in breadth ; generally, however, the longer ova are also broader ; thus, a batch of ova averaging 0·151 millim. in length, were on the average 0·088 millim. in breadth, while a second lot, averaging 0·186 millim. in length averaged 0·11 millim. in breadth. Thus the length varies to a certain extent independently of the breadth, giving rise to differences of shape as well as size.

The average length and breadth also, as may be seen from the last example, varied in the different batches.

The amount of variation differs, apparently, in the different species, eighteen ova of *Lepas anatifera* varying between 0·132 and 0·189 millim. in length, and between 0·101 and 0·12 millim. in breadth ; while seventeen ova of *Conchoderma virgata* varied between 0·158 and 0·186 millim. in length, and between 0·095 and 0·132 millim. in breadth.

MEASUREMENTS of the Ova of Cirripedes in parts of a millimetre.

	Average		Maximum		Minimum		Ripe embryo. Average		Nauplius. I	Nauplius. II
	Length.	Breadth.	Length.	Breadth.	Length.	Breadth.	Length.	Breadth.	Length.	Length.
Lepas anatifera . .	0·166	0·113	0·189	0·120	0·132	0·101	0·25	0·045	0·25	0·79
" *pectinata* . .	0·164	0·107	0·17	0·113	0·158	0·095	...	...	0·26	0·66
" *fascicularis*[1] .	0·155	0·09	0·17	0·095	0·145	0·082	...	...	0·35	0·6
Conchoderma virgata .	0·17	0·12	0·186	0·132	0·158	0·095	...	...	0·29	0·8
Dichelaspis Darwinii .	0·16[2]	0·08[2]	...	...	...	...	0·19	...	0·22	0·66
Otianmalus stellatus .	0·17	0·09	0·184	0·095	0·158	0·084	0·24	0·095	0·22	0·32
Balanus perforatus .	0·188[3]	0·107[3]	0·202	0·120	0·139	0·082	0·165	0·031	0·28	0·46
" *improvisus*[4] .	0·115	0·089	0·12	0·09	0·11	0·089	0·3	0·15	0·18	0·24
" *balanoides*[4] .	0·3	0·175	0·31	0·19	0·29	0·16	...	...	0·36	0·45

1. WILLEMOES-SUHM gives 0·26 as the length of the ovum, but I suspect from the above measurements (which, however, were made on spirit specimens) that the 2 is a misprint for 1.

2. FILIPPI. The measurements of the Nauplii are WILLEMOES-SUHM's.

3. HOEK gives 0·18 × 0·11 for this form; while LANG's (32) 0·5 × 0·3 is much too great.

4. MÜNTER and BUCHHOLZ.

5. HOEK.

In *Balanus improvisus* and *B. balanoides* the average measurements given are the means of the measurements of BUCHHOLZ and HOEK respectively.

PART II.—EMBRYONIC DEVELOPMENT.

STAGE A.

The freshly laid Ovum.—Polar Bodies.—Formation of the First Blastomere.

The minute changes undergone by the Cirripede ovum during the earlier part of the embryonic period of development have been studied, in more or less detail, but often with very divergent results, by FILIPPI (17), FRITZ MÜLLER (16), MUNTER and BUCHHOLZ (22), BOVALLIUS (26), WILLEMOES-SUHM (28), HOEK (30), LANG (32), NASONOV (40), WEISMANN and ISHIKAWA (43), SOLGER (49), and NUSSBAUM (44, 48, 51). Of these WEISMANN and ISHIKAWA, and SOLGER have confined themselves to the origin of the polar bodies. F. MÜLLER in *Tetraclita;* MUNTER and BUCHHOLZ, BOVALLIUS, HOEK, LANG, and NASONOV in *Balanus;* and WILLEMOES-SUHM in *Lepas,* have described the segmentation as total. FILIPPI alone denied the existence of a nucleus in the " nutritive sphere." The views of these authors will be referred to again in the sequel, and I may say that my results while agreeing in many particulars with the accounts given, yet differ considerably in some respects, and I shall endeavour to show that in Cirripedes we have an extreme case of the process known as epibolic gastrulation.

Lepas anatifera.

The freshly laid ovum of *Lepas anatifera* (fig. 1) consists of a finely granular dark protoplasm, coarse and fine yolk granules, together with oil globules of various sizes. A colouring matter (soluble in spirit) gives the ovum a characteristic blue colour in many of the species. It colours both the oil globules and yolk granules : it apparently does not exist in a solid form, but as a solution which may sometimes be seen as a transparent liquid between the viteiline membrane and the embryo. Thus in one example the embryo lay in a blue solution inside the vitelline membrane. The blue colour becomes in older embryos replaced by red. As the embryo grows older the coloured yolk is used up in the formation of the almost colourless or darkly granular protoplasm, and the colour of the egg accordingly becomes fainter and fainter. This remark also applies to the ova of other genera with colouring matter of a different tint.

The protoplasm itself consists of very small refractive granules embedded in a relatively small quantity of a clear hyaline substance which can rarely be distinctly seen except at the periphery of the ovum, in the pseudopodia given off at the anterior pole, or in the vicinity of the nuclei.

The oil globules in the Cirripede ovum vary considerably in size, number, and colour. In the species under consideration, they are relatively large and few in number, and often measure as much as 0·015 millim. or more. They may be readily recognized by blackening in osmic acid. In spirit specimens they are represented by cavities, and the circular spaces seen in thin slices are sections of these. I do not believe true vacuoles occur in the ovum of any of the Cirripedes examined.

The yolk granules (which vary in size with the species) commonly measure about 0·005 millim., but smaller sized ones occur between these, and such occasionally predominate. Under the influence of certain reagents they fuse ; in other cases they appear spherical, but when well preserved and numerous they are squeezed by mutual pressure into polyhedral bodies of fairly uniform size (in the meshes of which are a number of smaller rounded granules). A peculiarity of the yolk granules is that they are hollow, being perhaps comparable in this respect to the "pseudo-cells" in *Hydra* (34). The yolk parts with picric acid and takes up staining matter with much more difficulty than the protoplasm.

A nucleus is not visible in the ovum at this stage without special treatment.

The freshly laid ovum (fig. 1) takes a definite shape (varying with the species), which is nearly ellipsoidal in *Lepas anatifera*. The egg membrane seen round all advanced embryos is not yet present : the ovum is motionless, and the constituents are uniformly scattered over it.

The first change perceptible in the egg of *Lepas* consists in the slight elongation, and in the appearance of a clear spot at the point situated most anteriorly, *i.e.*, at the apex of the blunt end of the ovum. A small dome-shaped mass of clear protoplasm elevates itself above the contour of the egg, gradually becomes conical, bacillus-shaped, pear-shaped, and finally constricts off as a clear spherical vesicle containing a few refractive granules (fig. 2), and showing on staining a nucleus with several chromatin, elements.* This is the *first polar body*. It lies generally at the apex, but sometimes a little on one side, as has been stated by NASONOV for *Balanus improvisus* (40, 41).

Sections of the ovum just before the formation of the vesicle always show a nucleus, usually very small, situated peripherally ; this may be spindle-shaped or spherical according as division is taking place or not. The direction of the spindle is generally parallel to the longer axis of the egg, but in some cases was nearly parallel to the surface and to a transverse axis. The perfectly uniform distribution of the protoplasm and yolk granules is very evident in these sections, which stain with a uniform pale tint. No radial striations of the protoplasm round the ends of the spindle could be detected during the formation of either the first or second polar body.

The first directive spindle has already been observed by WEISMANN and ISHIKAWA (43). In this, as in other cases, the germinal vesicle changed into the first directive spindle while the eggs were still in the ovary.

---

* The number of chromatic elements varies ; it appears to be commonly four or five, but, in some cases was, as far as I could make out, as many as ten or twelve.

As the constriction off of the first polar body takes place, the external hyaline layer of protoplasm secretes a delicate firm pellicle—the vitelline or perivitelline membrane, which encloses the ovum until the Nauplius is hatched.   It was termed the "decidua" by FILIPPI, but was regarded by NUSSBAUM in *Pollicipes* as a vitelline membrane ; the correctness of this latter designation follows from SOLGER's observations on *Balanus*, as well as from my own investigations on *Lepas, Conchoderma*, and *Chthamalus*.

The fertilization must, as NUSSBAUM supposes in *Pollicipes*, take place before the next following stage, when the second polar body is formed within the vitelline membrane, ova found at this latter stage developing normally.   I can affirm the same of *Lepas*, with the addition that ova do not appear to undergo any further changes unless they have been fertilized.   In the case of the individual spoken of under the heading of fertilization, the ovigerous lamella on one side formed a soft mass, the ova of which were giving off the second polar body ; these later formed a perfectly normal blastoderm, and had evidently been fertilized.   On the other side, the ova were only commencing to issue from the mouth of the oviduct, and had not yet given off the first polar body.   These, which were carefully observed under the same conditions as the ova on the other side, soon gave off the first polar body quite normally, and formed a vitelline membrane, but proceeded no further.

From these facts we may, I think, conclude firstly, that the entrance of the spermatozoon takes place after and independently of the formation of the first polar body, but before the formation of the second ; secondly, that fertilization probably takes place in the mantle-cavity outside the oviduct ; and thirdly, that the formation of the vitelline membrane is associated with the origin of the first polar body, and quite independent of the act of fertilization.   This fact gains in interest when it is understood that a similar membrane is also probably associated with the formation of the second polar body in the case of *Peltogaster*, the early development of which I hope to give in a future communication.

The vitelline membrane preserves more or less the shape of the ovum at the time of its formation.   In nearly all Cirripedes an anterior or larger, and a posterior or smaller pole are readily discernible from the first, though the difference between them becomes accentuated as development proceeds.   It appears thus that the anterior and posterior poles are already determined at this period.   It may be that the ovum, at first indifferent as to which of the poles of the elongated unfertilized ovum becomes anterior, only becomes orientated upon fertilization.   HOEK's observation on *Balanus balanoides* seems to point to the conclusion that fertilization takes place at the posterior end of the ovum, but even in this case it may be that the pole to be fertilized was predetermined.   I have only one observation bearing on this point ; in the unfertilized ovum of *Lepas anatifera* I could distinguish no difference between the two poles ; it must be noted, however, that the two poles show little difference in this species during any of the early embryonic phases.

The entrance of the spermatozoon has never been witnessed in Cirripedes.   HOEK observed what were apparently remnants of spermatozoa beneath the vitelline membrane at the posterior end of the ovum in *Balanus balanoides*, though the occurrence of such beneath the membrane is somewhat puzzling, considering the mode of formation of the latter.

Sections made of ova of *Lepas anatifera* before or shortly after the formation of the first polar body show the first directive spindle or a small round nucleus with several chromatin elements. Though I searched very closely I failed to determine with certainty any body which might represent the male pronucleus, though the spermatozoon had evidently penetrated the egg. The male pronucleus must be exceedingly small and easily overlooked, otherwise it would be necessary to conclude that the fusion of the two pronuclei takes place immediately after the first polar body is formed (in which case it would be very rarely detected in ova which had given off the first polar body) ; but this seems improbable, though traces of a male pronucleus were never found in sections at any later phase, even in ova where the second polar body was being or had just been given off, though many ova were examined, the preservation and staining of which, as far as the chromatin elements were concerned, were excellent.

After the formation of the vitelline membrane and fertilization, the contents of the egg contract considerably (figs. 3, 44) ; that this is a result of fertilization follows from the fact that, while fertilized eggs contract, unfertilized ones remain uncontracted.*

The protoplasm now commences to undergo marked rhythmical contractions (figs. 44, 45), and after a time again swells somewhat ; the clear area at the anterior end becomes amœboid, and throws out short blunt pseudopodia (fig. 3), which are as often retracted.

A blunt process, similar at first to these, gradually rises higher, and eventually becomes constricted off, and forms the *second polar body*† (figs. 4, 43).

Sections at this stage show either a nucleus, or the second directive spindle close to the surface (fig. 100), with its long axis parallel to that of the ovum.

The contractions of the ovum are now very marked, and give it very different appearances at different times (*cf. Lepas pectinata*, figs. 44, 45, 46). These movements are accompanied by a gradual redistribution of the material of the ovum, together with probably a production of new protoplasm from the yolk ; this results in the accumulation of protoplasm in the centre of the egg, and soon in the formation of a protoplasmic part at one end, as distinct from the yolk (fig. 5 ; also *cf.* figs. 45, 83). The first attempts of the protoplasm to assume a polar position are often

* Among the unfertilized ova which showed diminution in size were a few that did not, and had given off a second polar body within the vitelline membrane ; whether these were also unfertilized and formed an exception, or were fertilised by stray spermatozoa, I cannot say.

† In the case of the unfertilized ova, though no second polar body was observed, one was certainly being formed, though prevented from emerging by the close adherence of the egg to the vitelline membrane, owing to the absence of the contraction of the ovum following upon fertilization, for nearly all the sections showed the nucleus dividing again at the periphery of the ovum, precisely as in fig. 100.

somewhat crude, and the protoplasm forms an irregular readily-staining mass, some times with small outliers sharply separated from the feebly-staining yolk.

The polar bodies become pale and disintegrated, and the external one often gets washed away. The protoplasm is at last mainly collected at the anterior pole of the egg, and the yolk at the other (figs. 6, 7). The protoplasm is grey in colour and finely granular; the yolk includes nearly all the larger granules and oil globules, but contains some protoplasm, which is centrally situated. The inferior or yolk half of the egg generally forms more than half of the whole, but often (fig. 7) the protoplasmic portion is the larger. The nucleus, which, during the period at which the ovum was undergoing contraction, was small and situated peripherally and anteriorly, and was invisible without special preparation, now becomes larger, and appears as a definite clear spot (or as a not easily staining vesicle in section), sometimes accompanied by radial striation of the protoplasm, and often visible only on pressure (fig. 7): it simultaneously travels towards the middle of the protoplasm. The surface separating the protoplasmic half from the yolk commonly intersects the ovum in a perfect circle, and marks off what will form the first blastomere; though marked off from the yolk pretty sharply peripherally, the central portion of the latter has a protoplasmic mass continuous with that of the blastomere. The protoplasmic portion has almost universally been regarded as ectoderm, and the yolk as endoderm. There is, however, in nearly all cases, a single nucleus in the protoplasmic portion; the yolk at this stage is devoid of a special nucleus and is in no way comparable to an endoderm cell.

In the further development of the embryo the protoplasmic part supplies the nuclei, while the yolk provides most of the material for the protoplasm of the remainder of the ectoderm, as well as for that of the mesoderm and endoderm.

Very generally (as in all the other species) the line of separation of the protoplasm and yolk is almost accurately transverse, as described or figured by FILIPPI, MÜNTER and BUCHHOLZ, HOEK, and LANG, for other Cirripedes. In many cases, however, it lies at a small or considerable inclination to the longer axis of the ovum (fig. 10); we may thus get, as NUSSBAUM has observed in *Pollicipes*, all transitions from a nearly longitudinal to a transverse plane, arising by simple variation, without it being at all necessary to assume, as NUSSBAUM has done, a rotation of the plane. NUSSBAUM supposed that the plane of separation (his first cleavage plane) is parallel to the longer axis of the ovum, and later, rotates and becomes transverse. I carefully investigated the earlier stages from this point of view, and convinced myself that no rotation whatever occurs. In the position of the polar bodies my evidence is at variance with the statements of NUSSBAUM, and the facts are, I think, conclusive in favour of my own view. The polar bodies, it is admitted on all hands, are formed at the anterior pole; if rotation occurs the polar bodies which adhere closely to the protoplasm, must, as NUSSBAUM assumes, share in the movement, and in the case of ova with oblique basal planes should show an obliquely situated polar body. Fig. 48

(*Lepas pectinata*) shows quite clearly that this is not the case;* mostly, however, the polar bodies become rapidly lost, and I think the bodies supposed by NUSSBAUM to be polar bodies, and found by him in the furrow between the yolk and protoplasmic half, cannot be such. I have also frequently seen cases when the wall was accurately transverse, and the polar body situated apically (figs. 6 and 7). Lastly I have been able to watch the gradual formation of the protoplasmic half in a single ovum ; the line of junction in these cases was transverse from the first. NUSSBAUM must then, I think, look elsewhere for support for the theory (44) he has based on this supposed rotation.

The shape of the protoplasmic portion of the egg at this stage is generally ovoid, a central plug of greater or less extent reaching into the middle of the yolk, which thus fits like a thick-bottomed bowl on to the central mass ; it is this ovoid mass which meets the periphery of the ovum in a circle.

In favourable sections the nucleus may often be seen to be dividing in a more or less longitudinal direction. Fig. 101 shows the spindle of the segmentation-nucleus. As the first blastomere becomes constricted off, a transverse wall separates it from the yolk-cell, which receives into its central protoplasmic mass one of the daughter-nuclei of the segmentation-nucleus (fig. 102). Traces of the spindle are often seen after the formation of the wall, and appear to penetrate the latter (figs. 102, 103). Finally the two daughter-nuclei become quite separate (fig. 104).

Attention may be drawn to the fact that the axis of the spindle of the segmentation-nucleus is not at right angles to that of the second directive spindle; it is quite possible, however, that rotation of the nucleus has taken place in the meantime.

STAGE B.

(A.) *Formation of the Second Blastomere.*

Lepas anatifera.

The protoplasm of the first blastomere gives rise to a portion only of the ectoderm. It has been supposed by all previous observers (BUCHHOLZ, HOEK, LANG, NASONOV, NUSSBAUM) that the ectoderm of the Cirripede arises by extension and division of this cell only. LANG states that in *Balanus perforatus* the earlier stages are produced by division of this cell according to a definite law, while NASONOV gives in *Balanus* a definite law for the extension of the ectoderm over the whole of the yolk. For some time after studying the segmentation of the ovum I myself believed that these statements might be generally true, but signally failed in endeavouring to find the law according to which division proceeded, and before long discovered that the result was not merely owing to very considerable variation in the development, but also to a process which has been completely overlooked. Between the stage with one

---

* NUSSBAUM does not figure or describe any case of this sort.

blastoderm cell, and that with three, I could find no transitions, as also between many other stages. Allowing everything for variation I was unable to explain the facts. After much time wasted in these endeavours I found the true explanation. The second blastomere does not come from the first, but *from the yolk*; this at once gave the clue to the development, and further progress was easy.

The first external indications of the formation of the second blastomere are the gradual appearance of small granules over a definite area of the yolk to one side of the first cell (*cf.* figs. 50, 86, 88): these, from the first appearance of the patch, have a radial arrangement round a point situated near the periphery: the granules are seen to be arranged in long moniliform rows, between which are other smaller ones not showing any definite arrangement: between the rows are clear and thin lines meeting in a small central space; at intervals along the clear lines dilatations occur. As the granular protoplasm increases, the dark central granular part of the yolk comes to the surface, and the whole body of the newly forming cell gets sharply defined peripherally (fig. 9); finally, the fully-formed cell may project considerably (fig. 12) above the surface, clearly defined peripherally, but not sharply marked off centrally from the yolk.

The portion of protoplasm belonging to the yolk after the formation of the first blastomere may, from the first, be situated at the surface (fig. 103), in which case the smaller spindle is obliquely directed.

The protoplasm of the cell is clearly in part a replacement of and outgrowth from the yolk, and it is very evident that some of the finely granular protoplasm is formed at the expense of the coarse granules and oil globules of the yolk, since the former at first, in many cases, increases greatly in bulk, and occupies the position of the latter, which undergoes corresponding diminution: a portion, however, often very considerable in the case of the second blastomere, is formed from the central granular mass of the yolk, which, with the nucleus, has come to the surface. The colouring matter of the yolk thus transformed is simultaneously lost, so that a contrast in colour occurs between the greyish protoplasm and the coloured yolk.

The nucleus sometimes undergoes division, while the granular matter still occupies a central position in the yolk, and thus we get the appearance shown in fig. 8.

Sections at this stage showed that the two nuclei were certainly in some cases the result of division, so that they could not represent, as NUSSBAUM maintains, the two pronuclei; they are very rarely seen until the separation of yolk and blastoderm is far advanced, and are preceded by a stage with a single clear nucleus, which at first is not visible. The two pronuclei are probably small; the female pronucleus certainly is at first, and the clear spot appearing with the separation of the protoplasm is almost certainly the segmentation-nucleus. I will not, however, absolutely deny the possibility of such nuclei being in some cases respectively the female pronucleus, and the male pronucleus greatly grown since an earlier stage.

The nucleus of the second cell, at first sometimes scarcely discernible, as the centre of radiation previously mentioned, grows into a clearly defined round vesicle, in the neighbourhood of which the granules for a considerable time retain their radial arrangement (fig. 14).

The second cell usually makes its appearance on one side of the yolk immediately below the first one, against which it abuts, while the free part of the circumference is part of a circle or oval. When completely formed it may occupy only a portion of the side on which it appears, or may extend down as far as the posterior pole of the yolk, and take up quite half of the latter (cf. fig. 48). When the protoplasm and yolk are both dark, as is often the case in *Lepas*, the first and second blastomeres are, at first sight, only to be separated with difficulty, and such appearances must, I think, have given rise to the statement of LANG that one of the daughter-cells of the first blastomere in *Balanus* grows down on one side and divides off. On careful examination and rotation of the ovum, the two are quite clearly separated, as in fig. 12.

Stages in the origin of blastomeres, in the way above described, from the yolk have evidently been observed by LANG, NUSSBAUM, and NASONOV, though the first-named stated that the appearance indicated the position of the blastopore, while both the latter would apparently regard it simply as the nucleus of the larger cell. That it does not mark the position of the blastopore is evident from the fact that it does not coincide with the point of closure of the blastoderm in any of the forms I have studied ; moreover, similar cells are seen to arise in quite different positions at later stages, sometimes two or more at a time. That it is not a mere nucleus is easily seen by examination. KORSCHELT and HEIDER, in their excellent 'Lehrbuch der vergleichenden Entwicklungsgeschichte,' truly remark that, judging from NASONOV's figures, new cellular elements are formed from the yolk.

I did not succeed in ascertaining what determined the side of the ovum on which the second blastomere appeared, but that it was neither light nor gravity is evident, firstly, from the fact that light is practically cut off from the ova in many cases ; and secondly, that the ova in ovigerous lamellæ of *Chthamalus*, which were sufficiently transparent to examine without disturbing their relative position, showed that the position of the cell was apparently quite capricious.

It may be noted that when the first epiblastic cell was separated from the yolk by a plane inclined towards one side, the second appeared very generally on the opposite side (figs. 10, 12 ; cf. also figs. 48, 87). This would seem to point to the conclusion already drawn by NASONOV, on observing the inclination of the plane, that the bilateral symmetry of the adult is determined at this early period. I failed, however, to confirm his statement that the polar bodies are always a little to one side.

(B.) Growth of the Blastoderm over the Yolk.

Lepas anatifera.

To render the following account more easily understood, it may be stated that the yolk is gradually covered by cells arising in the same way as the second blastomere, except that the central dark mass of granular matter, at first present in the yolk, is mainly used up in the formation of the second cell, and a fresh formation of protoplasm necessary; these cells arise generally in contiguity with the previous formed cells; the latter, in the meantime, divide.

The blastoderm covering the yolk gradually extends from the anterior to the posterior pole; and the complete blastoderm at the close of the stage is the result both of the origin of cells from the yolk, and of the division of such cells.

I spent much time in following the details of this process, with the result that I have been able to establish very considerable variation. It may be of interest, as indicating the purely physiological meaning of many of the processes of cell-division and disposition in Cirripede embryos, to give some of the main lines of development.

It will be convenient to refer to the plane passing through the junction of the periphery of the first blastoderm cell with the yolk as the basal plane, and to have a term for the nucleus with its surrounding protoplasm while still wholly or partly in the yolk. LANG employs RÜCKERT's term *merocyte* for such a cell in the case of meroblastic ova, and it appears to me that this term may be conveniently used in the present case.

We may, therefore, speak of the emergence of a merocyte at the surface, and term the cells which cover the yolk blastomeres, whether they arise directly from the yolk or by division of the blastomeres formed earlier.

The species whose embryonic development has been studied by myself or others are: *Lepas anatifera, L. pectinata, L. Hillii, L. fascicularis, Conchoderma virgata, Dichelaspis Darwinii, Scalpellum vulgare, Pollicipes polymerus, Chthamalus stellatus, Tetraclita porosa, Balanus perforatus, B. improvisus,* and *B. balanoides.* Other species were investigated by BOVALLIUS, but they were not distinguished, and the account given by him differs so completely from my own that I have been unable to make use of his observations; but it is probable, from his figures and descriptions, that the development is similar to that of other forms, a probability which is reduced almost to a certainty by the fact that one of the species was *Balanus balanoides,* a form investigated also by HOEK, whose account can be readily brought into relation with my own.

In *Lepas anatifera* the basal plane may be transverse (figs. 8, 9) or oblique (figs. 10, 12).

After the emergence of the second merocyte (II.) the first blastoderm cell (I.) may divide into two cells placed symmetrically on each side of II. (fig. 14); or the plane of division may make any angle with that containing the nucleus of II., and the long axis of the egg (fig. 13).

A third merocyte (III.) may then emerge either to the right or left of II. (figs. 15, 16); while II. may sometimes divide in the same plane as I.

In a second series I. remains at first undivided, and II. divides into two cells (IIA. and IIB.) placed symmetrically on each side of I. (figs. 17, 17A). The third

T 2

merocyte (III.) then emerges generally immediately below II., while a fourth (IV.) may appear beneath I. on the side opposite to II. In the meantime I. divides transversely into a cell (IA.) near the apex and a more posteriorly situated one (IB.) situated on the side opposite II.

Other variations occur, but enough have been given to show that there is no constancy in the mode of growth of the blastoderm over the yolk.

In the early as in the later stages, the merocyte before emerging from the yolk may not uncommonly be seen to give rise by division to a second merocyte (fig. 17A), which may either produce a blastomere near the spot where it arose, or may pass to a more distant part of the yolk.

(C.) *Completion of the Blastoderm and closure of the Blastopore.*

Lepas anatifera, &c.

The exact details of the further development were not ascertained. With a larger number of cells the examination becomes increasingly difficult, both because the investigation of the number and position of the cells is very laborious, and because it is almost impossible to say in the latter stages how any given arrangement has arisen, owing to the double mode of origin of the blastomeres. As, therefore, the early stages had shown that there was so large an amount of variation in the development, I was unwilling to spend more time on the subject. The figures of the various species will give a sufficiently good idea of the further growth of the blastoderm, and a brief description, together with the explanations of the plates, is all that is necessary.

The further growth of the blastoderm takes place in precisely the same manner as in the earlier stages, i.e., by the emergence of merocytes from the yolk and the division of blastoderm cells. While the merocytes are emerging as blastomeres, which generally takes place at the edge of the blastoderm, though sometimes at a distance from it, the earliest formed cells divide up further, so that the largest cells are sometimes found at the periphery of the blastoderm, and the smaller nearer the centre.

As in the earlier stages, there is a distinct tendency in the more advanced stages towards bilateral symmetry in the disposition of the cells, and the side of the ovum on which the second merocyte emerged and which took the lead in growth usually maintains, I believe throughout development, its lead. The blastoderm is finally completed in most cases on what is probably the dorsal side of the embryo at a little distance from the posterior end of the egg. The position of the blastopore, however, varies somewhat; it is generally apparently close to the position later occupied by the anus, but not uncommonly it is terminal or even sometimes ventral ; in one case it was at a considerable distance in front of the end of the embryo.

With the exception of these tendencies I could detect no law in the mode of growth of the blastoderm ; the variation is so great that the process may be said to be irregular.

The method of closure of the blastopore is difficult to observe, and was not witnessed in *Lepas anatifera*, but I succeeded in seeing it once or twice in other species (see figures of *Balanus perforatus* and *Chthamalus stellatus*).

The end of the yolk projects out at one point as a small rounded elevation (figs. 21 and 22), often of somewhat irregular outline when seen from above. A merocyte appears in the centre of this (figs. 68, 71, 94, and 127), and fills up the gap between the surrounding cells, and finally emerges from the yolk as a blastomere.

Immediately after this stage the embryo in *Balanus perforatus* consists of a single layer of cells enclosing an undivided yolk (fig. 70). All trace of the blastopore is lost, and I failed to observe any such pit as that described by NUSSBAUM in *Pollicipes* (51) at a later stage, and can confidently affirm that no such exists immediately after the closure of the blastopore.

I am unable to say how many merocytes take part in the formation of the blastoderm ; in all probability the number is variable, but not large. As the ovum is often half covered when four or five have emerged, some such number as nine or ten may not be far from the mark.

The number of cells composing the blastoderm at the time of the closure of the blastopore is also variable ; some embryos (*Lepas pectinata*) in which the latter was completely closed showed less than twenty cells, while others with an open blastopore showed a far greater number (*cf.* fig. 69 of *Balanus perforatus*).

In the vast majority of cases the yolk does not commence dividing until it is completely covered by the blastoderm, but rarely division into two takes place (*cf.* fig. 96 in *Chthamalus stellatus*).

The mode of formation of the blastoderm of all the other species of Cirripedia investigated agrees, as far as could be observed, in all essential particulars with that given for *Lepas*, but it has been thought well to mention some of the chief points under the heading of each species to illustrate the great uniformity in the general mode of development throughout the group, and, at the same time, to indicate a few points observed less perfectly in *Lepas anatifera*, as well as to call attention to slight variations not observed yet in this species.

(D.) *Stages (A), (B), and (C) in various species.*

Lepas pectinata.

In this as in all other species examined, the course of development is similar in all essential points. The ollowing points may be specially mentioned.

The egg is more ovate than that of *Lepas anatifera*, but does not differ otherwise to the eye. The colouring matter is blue, and the egg dark, the protoplasm and yolk not being readily separated. Two polar bodies were observed, the first being formed without, and the second within the vitelline membrane; it is worthy of note, in connection with NUSSBAUM's account of the earlier stages of segmentation, that the second polar body may be seen to be apically situated in eggs, the basal plane of which is transverse to the long axis (fig. 47).

The egg undergoes wave-like contractions (figs. 44–46), and the protoplasm collects at the broader end (fig. 46), and becomes separated off from the rest along a basal plane, which may be transverse (fig. 47), or oblique (fig. 48). This protoplasmic portion may be larger than the rest of the ovum (fig. 46), and the nucleus becomes visible after a time as a clear spot inside it (fig. 46).

The first blastomere (I.) may divide by a longitudinal constriction, or the second (II.) may do so first.   III. may appear beneath II., or on one side of it; IV. beneath I., and V. beneath III.   Sometimes I. divides into two cells placed anteriorly and posteriorly respectively.

Conchoderma virgata.

The egg is nearly ellipsoidal in this species : the colouring matter is blue; the protoplasm and yolk are both dark. The protoplasm collects at one end, the yolk at the other. The basal plane may be transverse or oblique. I. divides by a longitudinal constriction, the axial plane passing between its two daughter-cells making any angle with that passing through the nucleus of II. III. arises to the right or left of II., or II. divides, by a longitudinal constriction, into two cells, beneath which III. appears. The yolk becomes covered by blastomeres emerging as in *Lepas*; the blastopore occupies a sub-terminal (or terminal) position at the posterior end of the embryo.

Scalpellum obliquum.

In this species, according to HESSE, the colouring matter is yellow.

Pollicipes polymerus.

In this form, the development of which has been partially studied by NUSSBAUM (48, 51), the egg is ovate in shape. The first directive spindle, according to this observer, is formed while the ovum is still in the ovary. The first polar body is found outside, the second inside the vitelline membrane. The egg showed waves of constriction before the protoplasm and yolk have separated. NUSSBAUM describes two nuclei found in the ovum as the separation of the protoplasm and yolk progressed, and observing these, in some instances, to be in contact and in others remote from one another, regards them as the male and female pronuclei before and during union. The nuclei were at first not visible even on pressure; in other cases NUSSBAUM apparently only observed one nucleus, for he says "Einen Kern kann man nicht leicht übersehen ; eine kleine Spindel entzieht sich fast immer an frischen, durch Einlagerungen complicirten Eïinhalt der Beobachtung." It is evident then, that the facts are capable of another interpretation, namely, that the small segmentation-nucleus gradually enlarges and divides in two in a longitudinal direction as described above in *Lepas anatifera*, and I would apply the remarks made on page 137 also to the case of *Pollicipes*.

As in the species examined by myself, the basal plane may be transverse or oblique, and NUSSBAUM's theory that rotation of the plane takes place, has been already discussed on page 135. I. may arise by a longitudinal septum. The emergence of the second merocyte has evidently been seen by NUSSBAUM, but regarded, apparently, simply as indicating the nucleus of the large cell.

Chthamalus stellatus.

In this species the egg is ovate, the colouring matter orange, the yolk granules small (0·0025 millim.), and the oil globules about as large as in *Lepas anatifera*. It seems probable that light and gravitation have nothing to do with the orientation of the poles of the Cirripede ovum, since the ova in an ovigerous lamella are orientated in all directions, as may be seen in the case of *Chthamalus* by examining the edge of a thin lamella. The wave-like contractions accompanying separation of the proto-plasm and yolk are readily observed, and the protoplasm finally collects at one end (fig. 84), the nucleus appearing as a single clear spot (fig. 84). The basal plane may be transverse (fig. 84) or oblique (fig. 85). The polar body is terminal in ova with transverse basal planes (fig. 84), (see *Pollicipes*). I. divides transversely, its plane of division making any angle with the axial plane passing through II. (fig 88), generally a right angle. III. may appear to the right or to the left of II. (fig. 89), or may arise beneath II. which has divided before I. by a longitudinal constriction into two cells placed at the same level (fig. 90).

On one occasion a newly-formed blastomere (still, however, in connection with the yolk) was seen giving off a nucleus into the yolk. Fig. 93 represents this case. The nucleus in the yolk was in this specimen connected with that of the nearest blasto-mere, or rather with that of the protoplasmic mass which will become cut off in the next blastomere, by a spindle, seen by focussing below the surface. In the vast majority of cases, however, the division of the nucleus takes place before the merocyte forms any considerable prominence upon the surface of the egg.

Fig. 94 shows the blastopore becoming closed by the emergence of a blastomere from the yolk at this point.

Tetraclita porosa.

In *Tetraclita* the basal plane is figured by F. MÜLLER as transverse, and I. is divided into two daughter-cells placed side by side.

Balanus perforatus.

The ovum in this species is ovate,[*] the colouring matter is greenish-brown, the oil globules small and numerous, and the yolk granules (relatively to *Lepas* and *Chthamalus*) large (0·0075 millim.) ; their outlines form a polygonal network over the surface of the yolk. The first directive spindle was observed in *Balanus* by WEISMANN and ISHIKAWA (43), though the polar body itself was not seen. The protoplasm collects at one end (fig. 49), and is well contrasted in colour with the yolk. The basal plane, generally transverse (fig. 49), may be oblique. I. may divide by a longitudinal constriction, and III. may appear to the right or left of II. (figs. 53, 54). II. may divide in the same plane as I. (fig. 54), or into an anterior and posterior cell. In other cases II. may first divide into two daughter-cells placed side by side (figs. 55, 56, 57), while III. appears immediately beneath these, or close to the posterior pole of the embryo (figs. 55, 56) ; IV. then emerges beneath I. or elsewhere, I. having in the meantime divided into a more anterior and a more posterior cell (fig. 58).

Sometimes the two daughter-cells of II. divide by a longitudinal constriction each into two cells placed side by side (fig. 57). In many cases a merocyte may be' seen to divide before emerging from the yolk (figs. 53, 57, 63). Stages in the origin of the second blastomere from the yolk have evidently been observed by LANG, who regarded, however, the emerging merocyte as indicating the point of completion of the blastoderm ; that this is not so is evident from the fact that it does not coincide in position with the point of completion of the blastoderm in this form or any other examined.

Fig. 69 shows the blastopore which becomes, later, closed by an emerging merocyte (figs. 68, 71). Fig. 127 is a section passing through the blastopore at this stage.

Balanus balanoides.

The process of formation of the blastoderm in this species (see HOEK (30)) are probably quite similar to that described in *Lepas anatifera*, *Chthamalus stellatus*, *Balanus perforatus*, &c. The second polar body was seen by HOEK, but supposed by WEISMANN and ISHIKAWA to be the first. The basal plane is described and figured as transverse.

Balanus improvisus.

In this species (41, 49) the first directive spindle is formed while the ovum is still within the ovary. The first polar body is formed outside, the second inside the vitelline membrane. The basal plane here may be transverse or oblique, and II. may apparently divide by a longitudinal plane into two daughter-cells placed side by side.

---

[*] It may be noted that the ovate shape is apparently an adaptation to that of the Nauplius shortly before hatching, it being an advantage that the vitelline membrane should offer no serious hindrance to growth.

Dichelaspis Darwinii.

In *Dichelaspis* the colouring matter, according to FILIPPI (17), is vermilion. The basal plane is figured as transverse, and the first blastomere is divided into two cells placed side by side.

FILIPPI pointed out that no nucleus was visible in the yolk (his "nutritive sphere").

Scalpellum obliquum.

The colouring matter in this form is apparently yellow (12). The basal plane is figured as oblique, and I. divides in a longitudinal plane.

Lepas, Chthamalus, Balanus.

A study of the eggs of *Lepas anatifera, Balanus perforatus,* and *Chthamalus stellatus,* by sections, confirms that of the living ova, and furnishes evidence as to the part played by the nuclei.

Sections of *Lepas,* taken at the stage in fig. 8, and slightly later, show two nuclei in the newly-formed blastoderm; NUSSBAUM figures a similar stage in *Pollicipes,* but regards the two nuclei as the male and female pronuclei about to fuse. I have already stated my belief that these are merely the daughter-nuclei of the segmentation nucleus; one remains as the nucleus of the first blastomere, the other passes into the yolk hemisphere (figs. 102, 103, 104), where it transforms yolk material into protoplasm; the second merocyte formed partly in this way and partly from previously existing protoplasm, issues as the second blastomere, while the first becomes simultaneously cut off from the yolk (figs. 105 and 123).

Sections of eggs of *Balanus perforatus* show that the nucleus of the third merocyte is derived from that of the second (figs. 124*a* and 124*b*) ; the latter becomes spindle-shaped, and gives off a nucleus, which, accompanied by little or by no appreciable quantity of protoplasm, passes into the yolk, here it surrounds itself with protoplasm at the expense of the latter, and emerges generally close to the second blastomere; after division of the nucleus (fig. 106) the latter becomes separated from the yolk by a well-defined wall (fig. 125).

The third merocyte, in similar manner, while emerging as a blastomere, divides and gives off a nucleus to the yolk, which in a similar manner gives rise to new merocytes and blastomeres.

In the later phases of segmentating eggs of *Balanus,* stages are found in which no nuclei can be detected among the yolk granules, even upon the most rigid search, but in which, as in figs. 123, 124, and 125, the newly-forming blastomere is still in communication with the yolk (fig. 126) ; in others a nucleus with a large or small quantity of protoplasm (merocyte) is found in the yolk near the edge of the blastoderm.

Such nuclei are never found beneath the cells at the anterior pole of the embryo, which are already cut off from the yolk.

These facts, together with those relating to the division of emerging merocytes, given on pages 139 and 143, indicate that as the blastoderm grows over the un-nucleated yolk, the merocytes, before or during emergence, at its edge give off in turn nuclei into, and become successively cut off from, the yolk.   Once cut off they have no further connection with the yolk, which simply acts as an appendage and reservoir to the newly forming and growing cell or cells.   The nucleus, at first usually small, is generally in connection with little protoplasm, but soon transforms the surrounding yolk granules and oil globules into granular protoplasm with which it clothes itself, and in this way forms a merocyte ; this, at first in close proximity to the blastomere from which it has been formed, often becomes isolated in the yolk and forms a rounded protoplasmic nucleated body, with rays extending into the yolk in which it moves, and the yolk granules of which it probably devours, more or less after the manner of a phagocyte ; the emergence of such merocytes gives rise to blastomeres, which in their turn give off nuclei into the yolk before becoming cut off. In fig. 125 a merocyte is seen close to one of the blastomeres, and the complete series of sections shows that the nucleus is very small and the merocyte nowhere at the surface.   In figs. 126 and 127 a merocyte has come to the surface, but the cell formed has not yet been cut off from the yolk.

It is to be noted that the yolk is at no period cut off from communication with nucleated protoplasmic material in the way sometimes supposed in ova with much yolk material.

The division of the blastomeres and of the merocytes in the yolk is always accompanied by characteristic karyokinetic figures, which are often readily seen in surface views or in sections ; the radial arrangement of the protoplasm is often visible long after the spindle has disappeared.[*]

STAGE C.

(A.) *Formation and Division of the Meso-hypoblast, and Origin from it of the Mesoblast and Hypoblast of the Nauplius.*

The further stages of development are so uniform that, with the exception of the unimportant difference mentioned on p. 149, the following account of the later stages of embryonic development, though referring chiefly to *Lepas anatifera* and *Balanus perforatus* (of which alone sections were made), will, as far as known, apply to any one of the species.

The closure of the blastopore is almost immediately followed by the division of the yolk into two pyramids or segments ; the formation of the mesoblast immediately

---

[*] KNIPOVICHA has recently given (51A) a few details as to the development of *Laura;* he regards the segmentation as superficial, but it may in all probability be reduced to the type described in the present paper.—[13/7/93.]

commences by the successive cutting off and sub-division of nucleated segments from the two yolk segments.

It is very rare to find stages in which the blastopore is closed and these processes have not commenced; but by prolonged search cases may be found in which the yolk is still undivided (figs. 23, 70).

In these the blastoderm consists of a single layer of cells throughout its extent, one or possibly occasionally more cells, including that filling up the blastopore, being at first still in communication with the yolk, which, as shown by sections, as yet contains no nucleus apart from that of the protoplasmic mass with which it is in communication.

In other cases the blastoderm is completely cut off from the yolk, while a single large nucleus is found in the yolk at the posterior end surrounded by a protoplasmic body sending rays between the yolk granules (figs. 70, 128).

The yolk is now in the condition of a single cell; this will give rise to the hypoblast and mesoblast; in the vast majority of cases, however, the nucleus has divided into two, and with it the whole of the yolk (figs. 24, 72, 95, 96).

It seems à priori extremely probable that the nucleus is derived from that of the merocyte which filled up the blastopore; in rare cases, however, as NUSSBAUM has already observed in *Pollicipes*, the yolk apparently divides before closure of the blastopore; it is, however, difficult to make certain of this, as the last formed cells covering the blastopore and its neighbourhood are often very low and inconspicuous, and may be readily overlooked; but if, as I think, the blastopore is really open at this stage, the single nucleus in this case must be derived from another of the cells at the posterior end of the embryo. It is not perfectly clear, therefore, whether the last point left uncovered by the blastoderm coincides in all cases with the point of origin of the endoderm and mesoderm.

Immediately after the division of the yolk the posterior ends of the two cells now composing it become more opaque, owing to the replacement of the yolk granules by polyhedral cells about the same size as those of the epiblast (fig. 97). This process may go on till quite a considerable portion of each yolk-cell is transformed into cellular material. When it is complete the anterior uninvaded portion of the yolk, still consisting of two yolk-cells, will give rise to the hypoblast, while the cells formed at the posterior end constitute the mesoblast.

Sections of *Lepas anatifera* and *Balanus perforatus* (figs. 107 to 111, 129, 130), at the stages just considered, show clearly the manner in which this is brought about. The complete external layer of cells, which may now be termed epiblast, will give rise to the ectoderm of the adult. It consists of a single layer of rounded or flattened epithelial cells, often higher anteriorly and posteriorly, the nuclei of which usually occupy about the centre of the cells; over the whole of the embryo these cells are commonly elongated and rapidly dividing by radial, but never by tangential walls. Occasionally (figs. 107, 108) cases may be seen in which the nuclei might be supposed to be dividing along a radial plane, but careful comparison of adjacent sections shows that in this case the cells are cut more or less tangentially, and no proliferation of epiblast can be detected even upon the most rigid search.

I did not succeed in obtaining sections of embryos of *Lepas* in which no mesoblastic cells were formed, but in *Balanus*, figs. 129 and 130 represent such stages. The next stage seen is represented in figs. 107 and 108, where immediately below the epiblast are two elongated cells lying next to the two nucleated protoplasmic masses, which are continuous with the yolk. The two protoplasmic bodies are the essential parts of the two yolk-cells, and the two cells beneath are the first mesoblastic cells which have been cut off from them. This is seen from the facts that their outer boundaries are continuous with the contour of the yolk : the two cells are not coextensive with cells of the epiblast, at their junction with the latter, but are, on the contrary, coextensive with the yolk-cells, the nuclei of which may sometimes be seen dividing : they are also of a different character to the epiblast cells, inasmuch as they often contain yolk granules and oil drops, undigested remnants inherited from the yolk, and consequently the staining is less deep.

The growth of the protoplasm of the yolk-cells at the expense of the yolk continues, and the two nuclei with their surrounding protoplasm pass forwards ; new mesoblast cells are cut off from the two yolk-cells, while the earlier ones divide longitudinally and transversely (figs. 109 to 111). The origin of the mesoblast of the yolk-cells is frequently evident even when the cells are quite numerous, owing to the original contour of the yolk being often preserved ; to the delicacy of the last formed septa, which are found next to the dividing yolk-cells; to the proximity of the nuclei of the last formed cells to those of the yolk-cells; and to the difference in the depth of staining of mesoblast and epiblast. Gradually, however, the contents of the whole mesoblast cell becomes transformed into protoplasm, and the cells become indistinguishable except in position from those of the epiblast (fig. 111).

It may be noted that the mesoblast cells at first often seem to correspond in position on opposite sides of the septum between the two yolk-cells (fig. 110), but this correspondence soon becomes lost.

It is evident from this description that, upon closure of the blastopore, the whole yolk forms a simple cell, which will give rise to the mesoblast and hypoblast. This immediately divides into two yolk-cells, the protoplasmic portions of which are situated behind. From these two cells mesoblastic cells are successively cut off, which, like the epiblast, rapidly divide up (in a karyokinetic manner) and form a plug of mesoblast.

The mesoblast has been observed by NASONOV and by NUSSBAUM. The latter states that it is formed by the active division of superficial blastoderm cells round the edge of the blastopore before this has closed : but no such pit as that described by this observer in *Pollicipes* as existing at a time when the yolk-endoderm cells (see *postea*) are tolerably numerous, occurs in any of the species investigated by myself, and in none of these species did the formation of the mesoblast commence till the blastopore was closed.

My own results are more in accordance with those of NASONOV, who describes the mesoblast as originating from two symmetrically placed cells of the endoderm ; the.endoderm (so-called) at this stage, however, consists of only two cells (yolk-cells), and these, as will be seen immediately, are not placed symmetrically.

The yolk, as just mentioned, is divided into two segments at this period ; the cleavage plane, in most cases where I could make certain, traverses the blastopore, or passes from a point in front of it downwards and forwards to the ventral side. It is by means of the direction of this plane that the orientation of the embryo can be first determined, and the relation of the anus to the blastopore rendered probable. It is very difficult to identify the former position of the blastopore after it has closed, but in certain cases the last-formed blastoderm cell can be recognized, either by the similarity of colour (blue in *Lepas* and *Conchoderma*) to the yolk, or by the contours of the yolk and last blastoderm cell being continuous. In rare cases division of the yolk occurs before closure of the blastopore ; in most of these cases the first cleavage plane passes through or close to the blastopore. This plane can be recognized in later stages of the embryo, and is always seen to cut the plane of bilateral symmetry at right angles. NASONOV states that it is longitudinal ; it is, I believe, never so in the species I investigated, but always inclined forwards and downwards. The inclination varies somewhat, and the plane may even appear to be nearly transverse. The yolk is thus divided into an antero-dorsal and a postero-ventral segment : each is completely cut off from the other and consists anteriorly of the usual elements of the yolk, while posteriorly is a nucleated protoplasmic section sending rays into the yolk ; there is a single large nucleus in each, close to the plane of separation, and which has evidently arisen from the single nucleus present in the preceding stage.

The yolk segments, after the separation of the mesoblast, may be termed either yolk-pyramids or yolk-endoderm cells, and will divide into a number of cells similar in character, each of which will later give rise to an endoderm cell. The two may be regarded as endoderm, but their yolk portions probably furnish nutritive material for the later growth of the mesoblast cells round the endoderm, and for the elongation of the proctodæum and stomodæum.

The two yolk-pyramids remain for a time quiescent, and do not take part in the development. The epiblast and mesoblast cells divide rapidly, the epiblast, more especially towards the poles, become more columnar, and the mesoblast forms a considerable mass at the posterior end, truncating both of the yolk pyramids (fig. 25).

The epiblast cells at the posterior end in *Lepas anatifera* (but not in any of the other species) at this and rather later stages are often rounded, while those at the anterior end have flattened tops ; the posterior area thus limited, is often strongly intersected by certain lines (figs. 25–27) ; these, however, as far as I could make out, showed no constancy ; though they sometimes doubtfully corresponded with the

Nauplius segments seen at a later stage, at other times (figs. 25–27) they were more irregular.

The further development before the appearance of any definite organs consists in a multiplication of the epiblast cells, and a growth of the mesoblast in an anterior direction on the dorsal side.

(B.) *Extension of the Mesoblast.*

The mesoblast, at first concentrated chiefly at the posterior end of the embryo (fig. 112), grows forward at the expense of the yolk, and forms a thickish plate on the dorsal side obliquely inclined from the dorsal side in front to below the terminal point behind (figs. 131, 132). It gradually extends over the embryo ; on the dorsal side it is thick, but attenuates on the sides and is scarcely represented on the ventral side and in front. The peculiar position of the mesoblast is readily brought into relation with the development of other forms when it is understood that a considerable part of it will form the muscles of the appendages of the Nauplius, the latter arising first *on the dorsal side.* [*] .

By this extension of the mesoblast along the dorsal side, the yolk comes to appear nearer to the ventral than to the dorsal side, and this has given rise, I believe, to the statement that the ectoderm becomes thickened on this side.

The yolk-pyramids, which at first were quiescent, meanwhile divide generally by planes equally inclined to both sides of the plane of symmetry ; the division takes place in each of the two pyramids ; the nucleus and daughter-nuclei[†] divide tangentially in the plane of symmetry repeatedly into two : the division is never radial. We thus generally get two rows of yolk-pyramids, one dorsal, the other ventral, the nuclei of which seen from above or below lie in a single line (figs. 26–28, 71–75) ; in many cases, however, the division of the cells, though anticlinal, shows no relation to the plane of symmetry, and the resulting disposition of the pyramids is irregular, as HOEK has also observed in *Balanus balanoides.*

The number of yolk-endoderm cells present at this stage, varied from about two to eight in a large number of embryos of *Balanus perforatus* and *Lepas pectinata ;* sometimes the endoderm remained divided only into two during the whole phase.

The yolk-endoderm pyramids still form a solid mass, no archenteron being yet present.

---

[*] The sections of embryos of *Laura* (51 a, T. III., fig. 27), recently described by KNIPOVICHA, apparently also show a dorsal mesoblastic plate.—[13/7/93].

[†] These nuclei are usually very difficult to observe without the aid of sections in other species than *Balanus perforatus.*

STAGE D.

Formation of the Nauplius Segments.

This and the immediately following stages of the development, have been studied in more or less detail by FILIPPI, CLAPARÈDE, BUCHHOLZ, WILLEMOES-SUHM, HOEK, LANG, NASONOV, and NUSSBAUM.

Of these the observations of FILIPPI, CLAPARÈDE, WILLEMOES-SUHM, and NUSSBAUM are very brief.

FILIPPI (17) simply says that in *Dichelaspis* development begins on the ventral side.

CLAPARÈDE (14) confines himself to the statement that the segmentation of the embryo in *Lepas* is indicated by furrows.

WILLEMOES-SUHM (28) states that a groove appears, on each side of which the appendages arise. His figs. 5 and 6 do not, I am convinced, represent, as he supposes, an early stage in the segmentation of the ovum, but the division of the body, the blastoderm of which is already complete, and the endoderm well advanced in division, into the three segments of the embryo Nauplius. His fig. 7 similarly, I believe, indicates an embryo of a still earlier period, when the segmentation of the embryo had not yet commenced ; the internal cells he describes are evidently the yolk pyramids, and do not, as he supposes, represent cells which ever come to the surface to form the blastoderm.

NUSSBAUM (51) states that in *Pollicipes* the anterior part of the embryo is separated by a first furrow, followed later by a second posterior to it. The appendages he describes as arising on the three segments thus formed as simple protuberances which grow towards one another on the ventral side.

More details are given by HOEK, BUCHHOLZ, LANG, and NASONOV.

BUCHHOLZ (22) gives the most detailed description, but he failed to understand the development of the first pair of antennæ.

HOEK (30) observed the division of the body into three segments, on each of which arise a pair of protuberances which grow out and form the appendages.

NASONOV (40) describes a similar stage in which he, like BUCHHOLZ, observed a median longitudinal furrow.

All the above authors have regarded the surface on which the appendages appear first as ventral.

LANG, also, though he gives few details, took the ventral surface (see his fig. 20), with the labrum and tail for the originally-thickened side of the embryo, evidently regarding the side on which the appendages appeared as ventral.

In consequence of the unusual but simple mode of origin of the appendages, much confusion has arisen as to which is ventral and which the dorsal surface, and I shall give reasons for believing that, though certain stages have been correctly described and figured, these authors have transposed these two surfaces. The whole course of embryonic development has been thus thrown into great confusion, and it will, I think, be simplest if I describe my own observations, pointing out at the same time where I am in accordance with previous observers.

The first indication of the external organs consists in the appearance of two perfectly straight transverse furrows on the dorsal side of the embryo (figs. 29–31). These divide the body into three divisions, the relative size of which varies somewhat; generally the anterior one, which corresponds to the head and first segment, is largest. The most posterior division is generally a little smaller, but may be of a size equal to, or even larger than the anterior one; it corresponds to the third segment, and to the tail of the Nauplius together with the caudal spine. The middle division is always smallest, and corresponds to the second Nauplius segment.

The anterior and posterior divisions are simple, and show no indication of their composition, and there is no sign of a division of any of the segments into two halves.

The furrows are complete across the dorsal surface, and pass vertically downwards on to the sides where they die out, not extending to the ventral surface.

In sections (figs. 113 and 114) the yolk-endoderm nuclei are a little larger than those of the epiblast and mesoblast, and appear scattered about in the yolk, some being near the centre, others the periphery. The mesoblast and epiblast cells are closely associated, but the nuclei of the latter are generally elongated tangentially, rarely more or less radially. The mesoblast nuclei are a little larger than those of the ectoderm, and are generally elongated tangentially.

Sections at the posterior end show a solid mass of epiblast and mesoblast cells; a short distance in front there is a ring of these elements, the mesoblast being thickest dorsally, and enclosing an almost circular portion of yolk-endoderm. In the middle of the body the mesoderm is thickest dorsally, and, extending round the sides, dies out before reaching the middle line, being represented on the ventral side only by scattered cells (fig. 114).

The mesoblast in this stage is thus shaped rather like an inverted coal-shovel, being closed on all sides except in front and ventrally, where the cells are scattered or absent.

This phase must be of short duration, as it is not often met with.

STAGE E.

Marking out of the Nauplius Appendages.   (Figs. 32–34.)

The next change consists in the appearance of a median dorsal longitudinal furrow (fig. 34), which intersects all three divisions, terminating, however, before reaching either end. Simultaneously with this furrow two new transverse furrows appear, dividing the anterior and posterior Nauplius segments into two divisions.

The embryo is thus divided into an anterior median unpaired portion, three segments each defined by two transverse furrows, and divided into two symmetrical halves by the longitudinal furrow, and a posterior lobe.

The longitudinal furrow thus bounds internally what may be distinguished as the free ends of the appendages ($ant.^1$, $ant.^2$, $mnd.$) of the three segments.

The posterior unpaired division ($ta.$) represents the tail (thorax-abdomen) and caudal spine, the slight posterior division sometimes observed being the rudiments of the caudal forks.

Sections at this stage practically only differ from those of the preceding stage in showing the epiblast and mesoblast traversed by the dorsal longitudinal groove.

STAGE F.

Bifurcation of Second and Third Pairs of Appendages. Origin of Labrum, Œsophagus, and Intestine.

The furrows on the dorsal side of the embryo soon become more marked, and form considerable depressions (figs. 35, 36, 76).

This phase has been accurately described by MÜNTER and BUCHHOLZ (22).

NASONOV (40) also describes the longitudinal furrow and three of the transverse ones.

NUSSBAUM (51) states that the anterior division of the body is separated off by a first furrow, a second transverse one appearing only later; but this was not so in any of the forms studied by myself.

HOEK (30) speaks of two slight constrictions on the sides of the embryo as defining the appendages, but did not apparently trace them across the embryo.

LANG (32) describes two oblique furrows as indicating the division of the embryo into three segments; he has, however, I believe, confused the stage under consideration with a later one, and his figures, 17 to 20, refer only to the later stages, the constrictions seen in which separate the labrum and tail from an intermediate region.

In all cases the side on which the furrows appear, or towards which the free ends of the appendages are directed, and on which the mesoblastic plate lies, has been not unnaturally described as ventral. I have convinced myself, however, that this side is really dorsal; firstly, from the impossibility of reconciling the appearances presented by the various stages on the former hypothesis; secondly, because the free ends of the appendages are, in all cases I have observed, or which are figured by other observers, directed towards the dorsal side till a late date; and, thirdly, because it is on the opposite side that the labrum, mouth, and nervous system arise.

I found it difficult to believe that such views as those shown in figs. 77 and 98 were

MDCCCXCIV.—B.   x

dorsal ones, but have placed it beyond all doubt by tracing the two surfaces in all the species.

The error is one easy to make, and one I did not detect until I had followed the development in detail step by step.

Since the free ends of the appendages are closely applied in the middle line, if growth in length is to take place, they must move in a longitudinal direction. They begin, therefore, to take a more oblique direction, their axes being directed not merely dorsally, but also posteriorly (figs. 35, 36, 77 to 79, 98). The first pair remain simple throughout development, while the second and third begin to bifurcate (figs. 77, 98); the free ends are divided into two before the bifurcation can be seen in surface views, dissection often being necessary to show this.

Sections of *Lepas anatifera* and *Balanus perforatus*, taken when the appendages are still quite short, show no great difference from those of the preceding stage; the mesoblast, however, on the sides and ventral parts of the embryo is thicker (figs. 115, 133), apparently in places where the ventral muscles of the appendages will later form; dorsally, the free ends of the appendages may be seen, together with the portion of the now free dorsal surface, with its slight mesoblastic element between them (fig. 116).*

It is hardly correct to say with NASONOV (40) that most of the mesoblast goes to form the appendages; a not inconsiderable portion will form the muscles which, within the body, run to these appendages.

The boundaries of the mesoblastic cells are often hard to make out; their increase apparently continues to take place at the expense of the yolk, which they more or less replace, but from which they are always sharply marked off.

Sections of *Balanus perforatus* (the embryos of the same phase in *Lepas anatifera* were unfortunately lost) show essentially the same relations, but the *œsophagus* is now seen arising apparently in connection with an epiblastic thickening not far from the anterior end and on the ventral side. It appears as a solid ingrowth projecting upwards and backwards into the yolk (*cf.* fig. 118).

No trace of the proctodæum is as yet visible.

These observations are in harmony with NASONOV's statement that in *Balanus* the stomodæum arises before the proctodæum. I could not, however, confirm his view that the œsophagus arises as a definite epiblastic invagination like that shown in his figs. 22 and 23 (No. 41). The condition figured appears to be a later one.

When the appendages have attained a certain length and degree of obliquity, and very short setæ have become visible at their tips, the ventral surface shows in side views two slight notches (fig. 37), often scarcely discernible in ventral views as narrow grooves.

---

* In *Laura* (51A), also, a similar ventral thickening takes place; sections are stated to show a complete investment of lower layer cells (mesoblast).—[13/7/93.]

The anterior of these is situated below the base of the second pair of antennæ, and marks off a slightly convex area in front, which will form the *labrum (lbr.)*.

The second notch is short, and marks off posteriorly a triangular area which represents the *tail (thorax-abdomen)*.

These become prominent in the latest phases of this stage (figs. 38–40, 80), the labrum becoming marked out by a parabolic furrow with two slight notches, and the triangular tail by a distinct transverse furrow; the appendages at the same time completely cover the dorsal surface of the tail.

Sections of embryos of *Lepas anatifera* and *Balanus perforatus* at this stage (figs. 117–120, 134, 135) show a further increase of the mesodermal tissue of the appendages and the muscles supplying them.   At the anterior end, in front of the œsophagus, is a flat bilobed plate of cells which gives rise to the *brain* of the Nauplius.   It is probably of epiblastic origin, but the embryos are so small, and the mesoblastic cells so closely applied to the very similar epiblast, that it is often quite impossible to say whether a given cell belongs to one or the other layer.   The bilobed supra-œsophageal ganglionic plate is apparently continuous behind with a mass of cells (figs. 119, 134), the front part of which is traversed by the œsophagus, and which lies dorsal to, and extends into and fills up the projection forming the labrum. It is exceedingly difficult to make out the structure of this mass, but it probably contains both epiblastic and mesoblastic elements, as it occupies the site of the future labrum, sub-œsophageal ganglion, circum-œsophageal connectives, and the space dorsal to the free part of the labrum (*cf.* fig. 121); a horizontal line (fig. 119) appears to mark the future position of this space.   At the sides the mass is continuous with the bases of the appendages.   The œsophagus consists of a single layer of cells, arranged round an axis, and sometimes already enclosing a lumen.   It has already, in some cases, the curve seen in the ripe Nauplius, sections sometimes traversing it twice.   It is difficult to trace as far as the ectoderm, as it plunges into a mass of cells very similar to, and in close contact with, the cells of the epiblast, but some of which are almost certainly of mesoblastic origin.   The intestine is present as a tubular mass of cells, projecting from the posterior end of the embryo far into the ventral part of the yolk, or along its under surface (figs. 120, 134).   It is generally one layer of cells thick.   I failed to trace it very definitely to the ectoderm, as it comes into close connection with a mass of mesoblastic cells filling up the tail (fig. 117).   The extension inwards takes place, I believe, by division of the cells at the anterior end of the intestine, and by continued growth at the expense of the yolk-endoderm, which, however, almost certainly supplies only the material for growth, the cells themselves arising by division of the intestinal cells behind.   Surrounding the posterior end of the intestine may be seen an accumulation of mesoblastic cells, bounding the yolk behind.   These may be often seen to accompany the intestine as a thin layer (fig. 135).   A similar layer surrounds the œsophagus (fig. 134).   These layers, which have apparently accompanied the œsophagus and intestine in their growth inwards, evidently represent the future circular muscles of the gut.

The yolk-endoderm cells, during stages E and F, are very inconstant in number, but usually vary between six and twelve; at the commencement of this period they are sometimes as few as two in number, and at the close are usually about twelve. The yolk-endoderm nuclei are larger and more irregular than the rounded ectoderm or mesoderm nuclei.

The ectoderm of the hinder somewhat vertically compressed part of the body is thin dorsally, ventrally, and at the sides (figs. 120 and 135). It usually separates from the endoderm and intestine ventrally, so as to leave a small space beneath the latter, occupied by reticular tissue (fig. 120).

STAGE G.

Appendages Long, with Short Setæ.   Origin of Body Cavity.

As the appendages lengthen (figs. 41 and 81) they become more and more parallel to the long axis of the body, the setæ become more distinct, and a number of those characteristic of the Nauplius when first hatched can soon be counted.

In the latter half of this period the embryos become more transparent, and the various internal organs can be distinguished, obscured only by trains of granular matter (*gr.mat.*) apparently associated with connective tissue cells.

The labrum (*lbr.*) becomes free behind and at the sides, and a considerable depression found immediately beneath it and behind its point of attachment leads into the mouth (fig. 121); this has apparently arisen by the excavation of, or splitting in, the solid mass of tissue found beneath the labrum during the last stage.

The tail (*ta.*) becomes elongated and more definitely bifid behind, where it shows two sharp points. It is separated distinctly from the shorter caudal spine (*ca.sp.*), the contour of which is continuous with that of the carapace.

The glands of the fronto-lateral horns (*frl.gl.*), not distinguishable at the beginning of the period, become defined later as two clear spaces situated in the antero-lateral angles.

The brain (*br.*) begins to be visible externally as a bilobed mass situated at the anterior end in the middle line.

Sections of *Lepas anatifera* taken at this phase (figs. 121-122e) show the mesoderm cells of the appendages as elongated spindle-shaped cells, which will form the muscles, and which extend to the dorsal and ventral sides of the body. The mesoderm of the legs and body is continuous with a portion in the labrum. The layer of cells immediately surrounding the œsophagus and intestine respectively have given rise to a single layer of spindle-shaped cells representing the muscular coats of these structures.

The nuclei of the ectoderm cells of the tail and caudal spine are now considerably enlarged, as are also the cells themselves, which are very numerous, and form a thick strand, filling up nearly the whole of the cavity of these organs. In accordance with the rapid growth of these parts the cells have become spindle-shaped. The spindle-shape is more marked in *Lepas* (fig. 121) than in *Balanus*. Over the front portion of the ventral surface of the tail, however, the cells are not elongated, but simply enlarged, and constitute, at any rate in *Balanus*, the rudiment of the ventral plate seen later in the free Nauplius.

A cavity (*b.c.*), a part of which beneath the intestine is generally visible at the last stage (F), begins to form by the separation of the ectoderm and endoderm, and by irregular cavities arising by separation from one another of the differentiating cells of the partial layer of mesoderm found between the body wall and gut, or in the appendages, labrum, &c. (figs. 121*a*, 121*c*, 122).

On each side of the embryo near the anterior end are found two large comma-shaped cells closely applied to one another, and the tails of which point backwards (figs. 121*e* and 121*f*). The protoplasm is finely granular and stains rather deeply, and the nuclei are large and granular, and provided with a single large nucleolus. These are the glands of the fronto-lateral horns in the earliest conditions in which I have been able to recognize them. They are closely applied to the cuticle of the fronto-lateral horns without the intervention of other cells, and are therefore in all probability greatly enlarged and specialized ectoderm cells.

The 'yolk-endoderm is still devoid of a cavity ; the cells have increased in number, and the nuclei are now mostly more or less peripheral.

The sub-œsophageal ganglion is now clearly visible as a bilobed mass of cells, continuous behind with the single layer of ectoderm cells of the ventral surface. It lies above the newly-formed space, dorsal to the free part of the labrum (figs. 121 and 121*d*). In front it is continuous with two circum-œsophageal connectives (fig. 121*c*, *c.o.c.*), themselves in intimate connection with the ectoderm.*

Owing to the increase in length of the appendages and general growth of the embryo the egg membrane at the end of this stage has increased much in length and often in breadth (fig. 41).

STAGE H.

Appendages with Long Setæ. Appearance of the Nauplius Eye. Excavation of the Mid-gut.

Finally (figs. 42, 82, 99), as the appendages attain their full length, the setæ become long and the body becomes more transparent than before ; the brain, fronto-lateral

---

* Supra- and sub-œsophageal ganglia and circum-œsophageal connectives have been recently described in the young Nauplius of *Laura*.—[13/7/93.]

glands, alimentary canal, with its muscles, become more clearly visible; the circular muscles on the proctodæum are specially conspicuous.

The muscles of the limbs have now elongated, and form distinct muscle fibres, traversing the greatly enlarged body cavity, which now extends to all parts of the body.

The fronto-lateral horns (*frl.h.*) are clear in specimens taken out of the egg membrane. They lie parallel to the body and legs, following pretty closely the lateral margins of the carapace, and terminate posteriorly in rounded extremities a little beneath it. I regret that in consequence of the minute size of the embryos I did not succeed in observing precisely how these horns arose.

The fronto-lateral glands are now mainly filled with clear spherules of the nature of a secretion; these are closely appressed to one another, and assume thus a polyhedral shape.

The labrum acquires a shape approaching that of the free Nauplius.

The eye (*Npl.eye*) arises as two oblong yellowish-brown rectangular patches, united together in the middle line like the two halves of an open book, each half being closely applied to the lobe of the brain of the corresponding side. These gradually darken and become red; the red soon becomes, for the most part, obscured by the development of a black pigment, but a reddish tinge can often be perceived in dissected specimens at much later stages.

The granular matter is now more definitely arranged in two symmetrical masses, the greater part of which (fig. 42) lies at the sides of the gut for its whole length. Anteriorly these give off a branch passing outside the glands of the fronto-lateral horns, and another passing inwards towards the brain; the bands are united posteriorly above the intestine, and in front above the anterior end of the stomach and in front of the brain. Their nature is difficult to make out, but they apparently represent connective tissue cells, which will form bands of a peculiar tissue, which later occupy the same positions.

In ventral and lateral views a few large flattened granular cells lie outside the muscular wall of the proctodæum and stomach on their lateral walls, and meet in the median ventral line (fig. 82). These may perhaps give rise to the longitudinal muscles occupying this position later.

The yolk-endoderm cells usually number about fifteen or sixteen (or less), the additional numbers being in this and the last stage given by transverse anticlinal divisions, this process sometimes taking place tolerably symmetrically on each side of the sagittal plane. It is to be noted that the division of these cells is not accompanied by any of the karyokinetic figures such as occur in the case of both ectoderm and mesoderm cells, the process being apparently direct.

At the commencement of this stage the yolk-endoderm still forms a solid mass, but, as the time approaches for hatching, the yolk-pyramids, forming, as already stated, a single layer, commence to separate from one another at the centre (fig. 122*a*); a cavity is thus formed, which is continuous with those formed at earlier phases in the œsophagus and intestine. The separation of the yolk-pyramids is mainly due to a diminution in size they undergo, owing, apparently, to the using up of their yolk-granules for growth of the other tissues. The successive stages in the formation of the cavity of the mid-gut are seen in figs. 122a to e. In the final condition (figs. 122a and 122e) a single layer of cubical or tabular yolk-endoderm cells surrounds a wide cavity.

Though the cavities and walls of the œsophagus, mid-gut, and intestine are all now continuous, the three regions are still sharply marked off from one another, the yolk-endoderm cells having a large, more or less stellate, nucleus, surrounded by yolk-granules, which stain with difficulty, and the smaller cells of the fore- and hind-gut having small rounded nuclei surrounded by finely granular protoplasm, which stains readily.

The proctodæum, at its junction with the mid-gut (fig. 122d), expands to form a shallow funnel, which completes the posterior wall of the stomach. The few cells of this funnel show vertical striations, and are evidently glandular. They are very possibly already functional, and may be emitting a secretion, which acts upon the yolk-granules of the mid-gut and assists in the formation of the cavity, as the latter is often filled by a brownish plasma, presumably derived by solution of the yolk-granules. The lumen of the intestine is nearly always much larger than that of the œsophagus at this stage, and it is possible that the cells are already busy in absorbing the nutriment extracted from the yolk-granules of the mid-gut.

It may be noted that at no stage in the development of Cirripedes has an organ comparable to the "dorsal organ" of many other Crustacea been found.

The Nauplii are now ready to hatch, and, when they get free from the loosened mesh-work of egg membranes, they emerge from the mantle opening in clouds with each rhythmical movement of the adult.

PART III.—THE FREE NAUPLIUS: FIRST TWO LARVAL STAGES.

(1.) HISTORICAL SKETCH.

Various stages of the Cirripede Nauplii have been seen and described by a number of observers, but the descriptions have been in many cases incomplete, or even erroneous; I have found it necessary, therefore, in some cases to give fresh ones.

In the following sketch I have endeavoured to trace the history of our knowledge of the earlier stages.

The Nauplii of Cirripedes were first seen in 1778 by MARTIN SLABBER (1), who observed them issuing in clouds from the shell of *Lepas fascicularis*. He gives a quite recognizable figure of the Nauplii, but regarded them as distinct forms (*Monoculus marinus*), serving as food for the Barnacle.

After the discovery by VAUGHAN THOMPSON (2) in 1830 of the Cypris stage of *Balanus*, GRAY (3) in 1833 observed the nearly ripe Nauplii of *Balanus perforatus*. He doubted THOMPSON's discovery, because of the great difference between the two larval forms, and curiously professed to see the form of the adult in the Nauplius.

BURMEISTER (4), in 1834, figured the Nauplius (and Cypris stage) of *Lepas fascicularis*; but I believe, from his description, that the Nauplii were still unhatched, and apparently not quite ready to hatch, since he states that an eye was not visible.

In 1835, THOMPSON (5) observed and figured the free Nauplii of *Conchoderma virgata* and *Lepas anatifera*. The general character of the appendages, the carapace, fronto-lateral horns, and caudal spine were described; and the peculiarities shown by the Nauplii before attaining the second stage seen, but referred to individual inability to pass on to the adult.

Of KOREN and DANIELSSEN's paper (6) on Cirripedian development I have only seen the statement (quoted by GERSTAECKER, in BRONN's 'Klassen'), that the larva of *Alepas* hatches with six legs, though it was evidently a Nauplius.

In 1843, GOODSIR (7) figured and briefly described free Nauplii seen to issue from the shell of *Balanus balanoides*. The first two stages of *Balanus balanoides* and the first stage of *Balanus tintinnabulum* were also figured. GOODSIR showed that *Balanus* must pass through two forms, viz., the Nauplius and Cypris stage.

In 1851, SPENCE BATE (8) observed and figured the Nauplii of *Balanus balanoides*, *B. porcatus*, *B. perforatus*, *Chthamalus stellatus*, and *Verruca Strömia*. He criticized GOODSIR's statement that the body showed segmentation, and observed the frontal filaments,* the labrum, and the spines on the last two pairs of limbs, and the forked abdomen, and noted that the tail was used as a rudder.

In the Monograph (1851) on the Lepadidæ (9) DARWIN gave a description of the Cirripede Nauplius, based chiefly upon the Nauplius of *Scalpellum vulgare*.

In the volume on the Balanidæ (10), which appeared in 1854, the same author gave further details based on the Nauplii of *Pyrgoma, Coronula, Platylepas, Alcippe*, and the other genera mentioned above, and drew attention to the uniformity in the development of the Thoracica. He observed that the Nauplius eye was composed of two halves, and also fixed the position of the anus.

MAX SCHULTZE (11), in 1853, observed in the heliotropic Nauplii of *Balanus* and *Chthamalus* the Nauplius eye, consisting of two halves, and resting directly on the brain.

KROHN (13), in 1860, figured a Balanid (*Balanus ?*) and a Lepad (*Lepas ?*) Nauplius, and correctly but briefly described and figured the brain and suboesophageal ganglia connected together on each side of the oesophagus in *Lepas anatifera*; he also recognized the alimentary canal, and drew attention to the fact that the anus was dorsal.

---

* These were also seen and figured by SLADDER.

In 1863, CLAPARÈDE (14) described the Nauplius of *Lepas anatifera*, evidently immediately after the first moult, and gave details (though incorrect) as to the number of bristles on the appendages.

In 1864, appeared FRITZ MÜLLER'S " Für DARWIN," in which the Nauplii of various groups of Entomostraca and of Prawns were compared and differences pointed out. A figure of the Nauplius of *Tetraclita porosa* is given, and the frontal filaments described as arising directly from the brain.

FILIPPI (17), in 1865, figured Nauplii of *Dichelaspis Darwinii* before and after moulting once. The form of the Nauplius is described as characteristic of the genus.

In 1866, GERSTAECKER (19) summarized the previously existing accounts of Cirripedian development (1—17), criticizing KOREN and DANIELSSEN'S account for *Alepas*.

In 1869, MÜNTER and BUCHHOLZ (22) gave the first detailed account of the structure of any Cirripede Nauplius (*Balanus improvisus*) we possess. The divisions of the alimentary canal including histological details (gut muscles and cells of stomach) were given ; the flexor of the tail and dorsal system of muscles to the appendages noted, and attention drawn to the differences between the Nauplii of different species of *Balanus*.

VON WILLEMOES-SUHM (28) gave in 1876 the only approximately complete account we possess of the history of any Cirripede (*Lepas fascicularis*) from the birth of the Nauplius to the fixed Barnacle. He showed the glandular function of the fronto-lateral horns.

In 1877, HOEK (30) gave a good account of the development of *Balanus balanoides*, one of the forms investigated with considerably different results by BOVALLIUS (26). Good figures of the Nauplius before and after moulting once were given. The mouth, placed by previous observers at the end of the labrum, was given its right position at the base. A lens described by a number of previous observers in connection with the Nauplius eye was shown not to exist in *Balanus*.

In the same year LANG (33) described the external character of the first two Nauplius forms of *Balanus perforatus*, and one of *Scalpellum* apparently shortly after the first moult. The characters of the appendages were given in greater detail than in previous descriptions.

SCHMIDTLEIN (37) in the same year, and LO BIANCO (46) in 1888, gave details as to the time of appearance of the Nauplii of various Cirripedes.

In 1890, GROOM and LOEB (48) gave an account of the influence of light on the movements of the Nauplii of certain Cirripedes.

In the same year NUSSBAUM (49) figured Nauplii of *Pollicipes polymerus* drawn apparently shortly after the first moult.

(2.) METHODS OF OBTAINING THE NAUPLII.

If the shells of the adult Cirripede are cut open, and the ovigerous lamellæ of a large number extracted, the lamellæ will be found to be at very different stages of development : some may be apparently fully developed. If these be placed in a watch glass the Nauplii, if quite ripe, will hatch out by thousands, and can generally be collected at the two points respectively nearest to and farthest from the light ; if not far from this stage they will hatch out in limited numbers after a time. According to my experience it is only from such advanced stages, as a rule, that one can hope to get Nauplii, since once the lamellæ are taken out of the shell they appear to make only a limited amount of progress, and after a time development ceases and the eggs disintegrate, though a continuous circulation of water and air, or either alone, be kept up. The parent appears to have some peculiar beneficial influence on the eggs.

Where a thick shell like that of Balanus perforatus has to be cut through the process becomes very tedious, and if the Nauplii are not wanted in large numbers they may generally be obtained by placing a considerable number of the shells in a vessel, and·taking after a time by means of a pipette probes of the water from the part nearest to or farthest from the light. The Nauplii latest hatched will generally be on the side to the light, and the earlier ones on the remote side.

In such small forms as Chthamalus stellatus large sheets of the cohering shells may be knocked off the rock by means of a chisel, and the lamellæ collected in great numbers ; if the ripest (generally recognizable by their paler colour) be selected a large number of Nauplii will be readily obtained.

The Lepads are not always to be obtained in such numbers as Balanus and Chthamalus, and the Nauplii are best obtained by placing full grown individuals in a large glass vessel, and noting and isolating the individuals which give off the clouds of Nauplii. These can generally be readily recognized, as the Nauplii thus hatched out are more sluggish than those of the Balanids, and remain for a longer time in the vicinity of the parent.

Fishing with the surface-net will furnish Nauplii of Balanus perforatus of the first moult, and often of other moults, probably the whole year round at Naples, and in late winter and in spring the sea of part of the S.W. coast of England (Plymouth), or that around Jersey, may swarm with Cirripede larvæ (Chthamalus, Balanus). In order to obtain Nauplii of Lepas fascicularis it is, according to WILLEMOES-SUHM (28), during the day time necessary to fish at some depth below the surface, as these Nauplii perform the daily descent to depths characteristic of the pelagic fauna as a whole, while at night they may be obtained in vast numbers at the surface itself.

The first moult is, as a rule, speedily accomplished, as WILLEMOES-SUHM found in Lepas fascicularis, in fact so speedily that in this species, as well as in Balanus perforatus, after half an hour or so it is often difficult to find any individuals which have not undergone it. It appears to take place more slowly in most Lepads, but the number of batches examined was too small for generalization.

In consequence of this circumstance it is only comparatively rarely that one meets in the *Auftrieb* (result of surface-net fishing) with Nauplii which have not moulted once. On the other hand the Nauplii of all the species kept in confinement, with exceedingly few exceptions, did not moult more than once. I could get those of *Balanus perforatus* alone, in some cases, to moult a second time. All my efforts to rear the Nauplii were futile. Though they readily fed on a variety of substances, they remained for days, or even for weeks, without undergoing any change other than an unfavourable one. I tried with large and with small quantities of water ; with water or air circulation, or both combined, or with still water ; with covered or uncovered vessels ; with vessels with a glass or sandy bottom ; in water free from or containing sea-weed ; in every case, however, without success. The larvæ performed their daily migrations in the vessels in continually diminishing numbers, and without perceptible growth.

Most other observers have met with a similar experience ; thus MÜNTER and BUCHHOLZ failed with *Balanus improvisus* ; MAX SCHULTZE with *Balanus balanoides*, and WILLEMOES-SUHM with *Lepas fascicularis* at precisely the same stage. SPENCE BATE makes similar statements with reference to the larvæ of the genera he studied (*Balanus, Chthamalus*, and *Verruca*). This period may accordingly be termed a " critical " one in the life-history of Cirripedes.

In consequence, probably, of this difficulty the majority of the researches on the embryology of Cirripedes only cover some part of the period between the early stages of segmentation of the ovum and the completion · of the first moult, or commence with the free or fixed Cypris stage.

I have found it convenient, therefore, to divide this subject into two sections, the first of which, forming the present paper, covers the former ground, a second will deal with the later stages.

(3.) GENERAL STRUCTURE OF THE CIRRIPEDE NAUPLIUS OF THE FIRST TWO STAGES.

The Nauplii of Cirripedes show marked differences from those of other Crustacea, yet the different forms, so far as I have examined them, or been able to ascertain from previous accounts, present great uniformity in their characters, and it will render the following comparative accounts more intelligible if I give a brief sketch of the main features which characterize the Nauplius during the first part of their period of existence (woodcuts, figs. 1 and 2).

The short and nearly colourless unsegmented body of the Nauplius is covered over a great part of the surface by a dorsal, more or less shield-shaped chitinous *carapace*, produced at the antero-lateral angles into short or long *fronto-lateral horns* (*frl.h.*), each of which is pierced after the first moult by an opening at the distal end for the passage of the secretion of the glands situated at their base. Two smaller glands also open at two other points on the margin of the carapace in some species. The anterior margin is nearly straight, or slightly convex, the lateral margins are more rounded. The main body of the carapace is indistinctly separated posteriorly by a faint transverse line of articulation from a long movable spine, covered with spinelets, and arising from a broader base, the sides of which are continuous with the lateral convex margins of the carapace. This is the *caudal spine*, and may be regarded with DARWIN as belonging to the carapace.

Fig. 1.

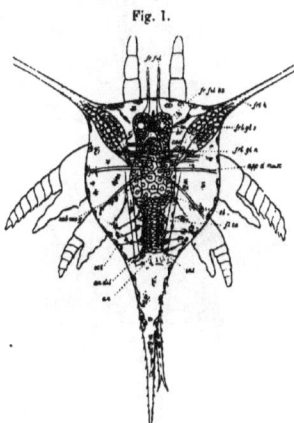

Diagram showing the characters common to Cirripede Nauplii after moulting once. The tail is represented as forked, though it is simple in some genera. The sub-œsophageal ganglion is hidden by the stomach, its posterior margin is indicated by a curved line, crossing the hinder part of the latter. The bristles and spines are omitted from the appendages in this figure. Dorsal view. (Lettering as in fig. 2.)

On the ventral surface, at a comparatively short distance behind the anterior margin, the very large *labrum (lbr.)* (with a small opening at the distal end for the passage of a glandular secretion) arises and projects downwards and backwards, making a considerable angle (fig. 151) with the body, and protecting the mouth, which lies immediately behind its base.

About two-fifths of the distance from the base of the labrum to the anterior margin spring two delicate transparent filiform organs—the *frontal filaments (fr.fil.)*—which are directed downwards and forwards beyond the anterior margin of the carapace.

At the sides of the base of the labrum, and in front of the mouth, spring the first pair of appendages or *antennules* (*ant.*[1]), composed of a single branch, and after the first moult very commonly directed forwards.

Fig. 2.

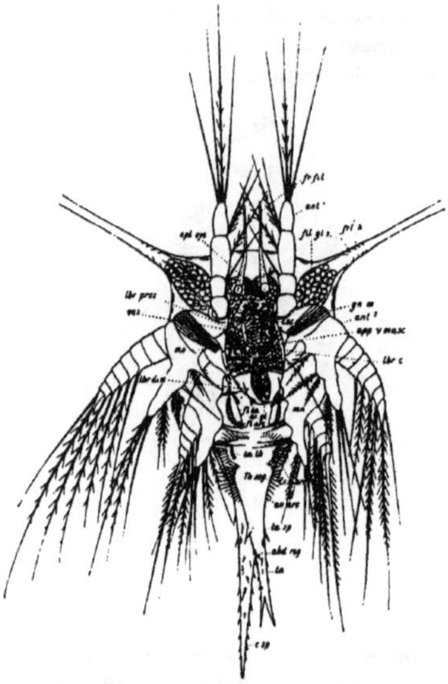

Ventral view of same. *an.*, anus; *an. arc*, anal arc; *abd. reg.*, abdominal region; *an. dil.*, dilator of anus; *ant.*[1], antennule; *ant.*[2], antenna; *app. d. musc.*, dorsal muscles to appendages; *app. v. musc.*, ventral ditto; *ax. gl.*, axial gland of labrum; *c. sp.*, caudal spine; *ect.*, ectoderm; *ex. m. arc.*, extra-maxillary arc; *fl. arc.*, flexor arc; *fl. ta.*, flexor muscle of tail; *fr. fil.*, frontal filament; *fr. fil. bs.*, base of ditto in brain; *frl. gl. n.*, nucleus of fronto-lateral gland; *frl. gl. s.* (*fil. gl. s.* in one figure), secreted spherules of fronto-lateral gland; *frl. h.*, fronto-lateral horn; *gn. m.*, muscle to gnathobase of antenna; *int.*, intestine; *lbr. c.*, cells at sides of labrum; *lbr. dist.*, distal lobe of labrum; *lbr. prox.*, proximal ditto; *mn.*, mandible; *mo.*, mouth; *npl. eye*, Nauplius eye; *œs.* (*ves.* in figures), œsophagus; *st.*, stomach; *sub-œs. g.* (*sub. ves. g.* in figure), sub-œsophageal ganglion (seen through the labrum); *ta.*, tail (thorax abdomen); *ta. sp.*, tail spine; *ta. th.*, tail-thickening or rudiment of "ventral plate."

Opposite the front corners of the mouth spring the second pair of appendages or *antennæ* (*ant.*[2]), the largest of the three pairs.

These are followed immediately by the third pair of appendages or *mandibles* (*mnd.*), which, like the second, are biramous.

The character of these appendages is remarkably uniform throughout the group, and will be given later on.

Behind the mouth is found a median region, traversed or bounded by characteristic bands of hairs, generally directed forwards, and forming with those of the labrum a sieve, which apparently serves to retain particles of any size which have found their way beneath the labrum. This region may be termed the *setose region;* it occupies the position of the future gnathites.

Posterior again to this is the *tail* (*ta.*) of the Nauplius, consisting of two regions, an anterior of short, truncated conical shape, extending from the setose region to the level of the two strong spines (*ta.sp.*), and corresponding in position with the future thorax : this may be termed the *thoracic region* (*th.reg.*) ; while the region marked anteriorly by the spines and forming a long spine covered with spinelets and generally bifurcated towards the end may be termed the *abdominal region*, since it marks out the region of the future abdomen and caudal appendages of the Cypris-stage.

Until after the completion of the first moult the tail and caudal spine are both telescoped.

The *mouth* (*mo.*), situated behind and at the base of the labrum, leads into a short *œsophagus* (*œs.*), which passes at first slightly downwards and forwards, and then curving round, at first upwards and forwards, and then upwards into the stomach. The *stomach* (*st.*) is a large simple rounded sac occupying the centre of the body. The *intestine* (*int.*) is straight, and about as long as the stomach, from the posterior end of which it passes to the anus. Like the œsophagus, it undergoes incessant peristaltic contractions. The *anus* (*an.*) lies at the base of the tail and on its dorsal side, that is between the tail and caudal spine.

The *nervous system* consists of, firstly, a lobed bilaterally symmetrical *brain* (*br.*), the halves of which are closely applied together in the middle line ; secondly, two stout *circum-œsophageal connectives* (*c.o.c.*) ; and thirdly, a *sub-œsophageal ganglion* (*sub-œs.g.*).

The *Nauplius eye* (*Npl.eye*) is black, and consists of two halves, sometimes separated, but generally closely applied together in the middle line, each resting on one of the halves of the brain.

On each side of this the brain gives direct origin to a frontal filament.

At the base of the fronto-lateral horns are two large spindle-shaped glandular organs—the *fronto-lateral glands* (*frl.gl.*)—each consisting, in all cases, of two closely applied unicellular glands attached on one hand to a point not far from the middle of the carapace above the sides of the stomach, and on the other opening into the cavity of the fronto-lateral horns.

Two other glands—the *lateral glands* (*lat.gl.*)—(probably also unicellular), usually much smaller, occur in some genera in the posterior part of the body, where they open externally by minute pores on the sides of the carapace.

The body cavity (using the term in the descriptive sense of a cavity included within the body-walls and surrounding the viscera) is spacious, and extends into every portion of the body. Its main section is bounded posteriorly by a delicate sheet of tissue (*an.dil.*) which separates off the cavities in the caudal spine and tail, and probably serves to dilate the anus. The body-cavity is transversed also by strands of branched and anastomosing connective tissue cells and contractile fibres, extending, at fairly frequent intervals, between the alimentary canal and body wall.

Distinctly *striated muscles (app.d.musc.* and *app.v.musc.*) also pass from the median dorsal and median ventral lines above and below the stomach to the appendages; others are limited to portions of the appendages.

With the exception of those of the walls of the alimentary canal the only other muscles to be found are the *flexors of the tail (fl,ta,)*, a pair of muscles which pass from the carapace (close to the attachment of the fronto-lateral glands) downwards and backwards to the sides of the setose region.

A special accumulation of mesoderm cells at the sides of the labrum will be referred to later.

On part of the ventral surface of the setose region in most of the genera, and in *Balanus* also in the thoracic and abdominal region, a special accumulation and disposition of the ectoderm cells (*ta.th.*) occurs on each side, the cells being at the same time peculiarly modified; the fate of these cells will be followed hereafter.

(4.) DETAILED STRUCTURE OF THE NAUPLII OF STAGES 1 AND 2.

Stage 1. Between the Periods of Hatching and of the First Moult.

The newly hatched Nauplii are active little creatures (figs. 140, 153, 161, 164, and 167), much smaller than after the completion of the first moult.

The dimensions in the different species are as follows: *Lepas anatifera*, 0·25 millim.: *L. pectinata*, 0·26 millim.; *L. fascicularis*, 0·35 millim. (28); *Conchoderma virgata*, 0·29 millim.; *Dichelaspis Darwinii* and *Chthamalus stellatus*, 0·22 millim.; *Balanus perforatus*, 0·28 millim.; *B. improvisus*, 0·18 millim. to 0·19 millim. (22); and *B. balanoides*, 0·36 millim. to 0·37 millim. (30).

The young Nauplius is more or less pear-shaped, or sometimes conical; broadest in *Balanus*, and narrowest in *Dichelaspis*.

The telescoping of the tail (*ta.*) and the posterior spine (*c.sp.*), and the direction of the fronto-lateral horns, are the most striking features of the Nauplius at this stage. The dorsal views show the telescoping of the dorsal spine, which lies invaginated within the body and extends as far as a point above the centre of the stomach. The point alone projects from the tube. The tail is similarly invaginated and projects as one or two points according as the abdomen is simple or forked.

The long or short fronto-lateral horns are directed backwards, extending in *Lepas* and *Conchoderma* nearly to the end of the body. Their ends are not open, as in all the later stages, but always closed and perfectly rounded. They are often seen to contain spherules of a glandular secretion.

The frontal filaments (*fr.fil.*), present in all the later Nauplius stages as well as in the Cypris stage, are absent, as some earlier observers have pointed out. I failed to find the stump-like rudiments described by LANG as representing them in *Balanus perforatus*, but could sometimes observe the filaments lying under the cuticle about to be cast off.

The appendages are only indistinctly articulated, and the hairs, spines, and teeth, which are present in considerable number, are all simple, *i.e.*, not plumose or furnished with secondary spines. In consequence of the indistinct articulation of the joints, and of the fineness and shortness of the hairs, both the number of joints and hairs were exceedingly difficult to ascertain, but the disposition given in the table on p. 182 certainly holds good for *Balanus perforatus* and, probably (from the perfect similarity in the appendages of all the species after the first moult), for the rest of the species. The agreement of Stage 1 with Stage 2 is seen from the table to be very close, the former having rather fewer joints and hairs.

The labrum has not yet acquired its definite shape, the distal lobe having rounded angles, and the teeth and hairs characteristic of it at later stages are absent.

The hairs on the setose region of later stages are also absent.

The granular tissue is plentiful and still prevents the internal organs being seen perfectly. It runs in fairly defined trains at the sides of the alimentary canal, above the anterior part of the stomach and posterior part of the brain, and in front of and behind the fronto-lateral glands.

In other respects than those just mentioned, the whole organization is essentially the same as after the first moult, when the structure can be made out more clearly.

Figures and descriptions of Cirripede Nauplii at Stage 1 have been already given for *Conchoderma virgata*, by THOMPSON (5); *Balanus tintinnabulum*, by GOODSIR (7); *B. balanoides*, by GOODSIR and SPENCE BATE (8); *B. perforatus*, by BATE and LANG (33); *B. porcatus, Verruca Strömia*, and *Chthamalus stellatus*, by BATE; *Dichelaspis Darwinii*, by FILIPPI (17); *Balanus improvisus*, by MÜNTER and BUCHHOLZ (22); *Lepas fascicularis*, by WILLEMOES-SUHM (28); and *Pollicipes polymerus*, by NUSSBAUM (51).

Few of these, however, are sufficiently detailed to be of much value for comparative purposes; the best figures are those of *Balanus balanoides*, by HOEK; *B. improvisus*, by MÜNTER and BUCHHOLZ; and *Lepas fascicularis*, by WILLEMOES-SUHM.

Stage 2. (Between the Periods of the First and Second Moults.) Figs. 210, 211, 220-222, 225-227, 232, 235, 238.

(A.) *The First Moult.*

Nauplii after the first moult have been figured for *Lepas anatifera*, by MARTIN SLABBER (1) and CLAPARÈDE (14); for *Lepas anatifera* and *Conchoderma virgata*, by VAUGHAN THOMPSON (5); for *Balanus balanoides*, by GOODSIR, SPENCE BATE, and HOEK (7, 8, 30); for *Verruca Strömia* and *Chthamalus stellatus*, by SPENCE BATE; for *Scalpellum vulgare*, by DARWIN and LANG (10, 32); for *Tetraclita porosa*, by FRITZ MÜLLER (16); for *Balanus perforatus*, by MÜNTER and BUCHHOLZ (22); for *Lepas fascicularis*, by WILLEMOES-SUHM (28); for *Balanus perforatus*, by LANG (32); and for *Pollicipes polymerus*, by NUSSBAUM (51). Of these the best figures are those of BUCHHOLZ, WILLEMOES-SUHM, and HOEK : those of LANG and SPENCE BATE are also valuable.* None of these authors, however, have given detailed accounts of the structure, and the descriptions are not always correct; nor have any comparative accounts been given of the Nauplii of the different genera and species. I have in the following account attempted to partially fill up this gap in our knowledge by means of a careful study of the Nauplii as seen whole, dissected, or in serial sections.

The increased transparency of the larvæ, due to the disappearance of the granular material, and to the expansion of parts crowded together beneath the cuticle of the Nauplius of the first stage, renders the Nauplius much more favourable for examination than before, while the transparency of the tissues beneath the carapace renders them better for study than the later Nauplius stages.

The time which elapses between birth and the first moult appears to vary according to conditions, and with the species. I have practically no recorded observations on this question, but according to the best of my recollection the Nauplii of *Balanus perforatus* at Naples underwent their moult within the first half-hour or so ; *Chthamalus stellatus* at Plymouth and *Lepas pectinata* at Naples were similar, and rapidly attained the full development of the second stage ; while *Lepas anatifera* and *Conchoderma virgata*, whilst possibly undergoing the moult with equal rapidity (though I did not observe this), were often many days before the tail, caudal spine, and appendages showed their full development, though sometimes the change was more rapid ; I suspect that in their natural condition the change is generally more rapid, as the movements of the comparatively sluggish Nauplii seemed to be retarded by a thin coating of viscid material which formed at the bottom of the dishes in which they were placed. Mere confinement under artificial conditions may also, very possibly, have some effect on the rate of change.

SPENCE BATE found in all the species he examined (Balanids) that the moult took

* I regret that I have been unable to see HESSE's second paper on the development of *Scalpellum* (25) in which figures, which may include the stage under consideration, have been given.

place on the second day : in *Balanus improvisus*, according to BUCHHOLZ, the moult took place at latest on the third day. GOODSIR found in *Balanus balanoides*, also examined by BATE, that the moult took place after eight days. It would appear possible that the duration of the first stage depends partly on latitude, but not altogether, since *Chthamalus stellatus*, one of the species examined by SPENCE BATE, and stated to moult on the second day, sometimes, as stated above, moults very shortly after hatching, so that if the observations are correct other conditions than latitude must come into consideration.

The changes accompanying the first moult were briefly characterized by WILLEMOES-SUHM in *Lepas fascicularis*, but have generally been ill-understood, and have given rise to some misconceptions. They are simple, however, and need but brief description.

When the time for moulting arrives the cuticle of the first stage is burst and thrown off. The Nauplius issues from a large gap at the anterior end of the cuticle ; the water entering this gap probably distends the cuticle, and assists in the process of exuviation which is apparently brought about by ordinary locomotive movements.

The tail and posterior spine, hitherto telescoped far within the body, become evaginated ; the hairs formed inside those of the previous stage were likewise telescoped, and as they become evaginated the secondary hairs with which they are furnished, and which at first must lie parallel to the sides of the tube inside which they were formed, become visible. Thus the tail, caudal spine, and bristles often appear jointed, the joint occurring where the invaginated portion begins (figs. 165 and 166). CLAPARÈDE has already seen this telescoping of the hairs in the Nauplii of *Lepas anatifera*, and LANG in those of *Scalpellum vulgare*, while WILLEMOES-SUHM observed it in the case of the tail and caudal spine of *Lepas fascicularis*, and HOEK in the case of the caudal spine in *Balanus balanoides*.*

This appearance, however, has given rise to the statement (33) that the caudal spine in *Balanus perforatus* consists of two segments.

WILLEMOES-SUHM has given the same explanation as myself.

NUSSBAUM, on the other hand, observed Nauplii of *Pollicipes* sometimes with a long, and at other times with a short caudal spine ; the shorter, he states (p. 29), is only the longer drawn within the body by a muscle, the spine being protruded or retracted at will. This statement is apparently based on inference, as one can readily perceive in *Lepas* and *Conchoderma* that the length of the tail and caudal spine depends on age ; the supposed muscle is simply a strand of somewhat elongated ectoderm and connective tissue cells.†

---

* GROBBEN (65) and CLAUS (61) figure the caudal bristles of *Branchipus* as arising from a kind of sac apparently in the same way as the setæ of Chætopods : I suspect, however, that they are simply telescoped as in the above-mentioned cases, and that the invaginated sacs have no essential resemblance to the permanent sacs of Chætopods with their special muscles. The telescoping of the tail and caudal spine suggests comparison with the method of formation of organs in Hexapod insects by "imaginal discs."

† The "peculiar chitinous structures" figured by KNIPOVICHA (51A) in the young Nauplius of *Laura*

The gradual evagination of these structures, and the curving forward of the fronto-lateral horns form the most marked changes the Nauplius undergoes at this moult ; the frontal filaments appear at once, as also do the bristles and teeth of the setose region and labrum, and the carapace becomes larger; the internal organization clearer.

The changes resulting in this transformation are so marked that LANG assumed two or three moults to be necessary for their completion in *Balanus perforatus*. In *Lepas* and *Conchoderma*, where the contrast is much greater owing to the length of the tail and caudal spine, seeing all stages of the transition, I thought that not less than four or five moults had taken place, and spent some time in fruitless endeavours to find constant differences between the different phases.

Several examples of these stages are figured (figs. 155, 165, 166) in *Lepas anatifera* and *Conchoderma virgata*.

LANG's figure of *Scalpellum* also probably represents a Nauplius at this transitional stage, as also do THOMPSON's figures of *Conchoderma virgata*, CLAPARÈDE's of *Lepas anatifera* and FILIPPI's of *Dichelaspis Darwinii.*

(B.) *Size of the Nauplii.*

The Nauplii, after moulting once, attain a size much greater than that of the first stage.

The lengths (taken in the species I examined from the front margin to the end of the caudal spine) are as follows :—*Balanus improvisus*, 0·24 millim. (22) ; *Balanus balanoides*, 0·45 millim. (30) ; *Balanus perforatus*, 0·46 millim. ; *Chthamalus stellatus*, 0·32 millim. ; *Dichelaspis Darwinii*, 0·66 millim. (17) ; *Lepas fascicularis*, 0·6 millim. (28) ; *Lepas pectinata*, 0·66 millim. ; *Lepas anatifera*, 0·79 millim., and *Conchoderma virgata*, 0·8 millim.

The Nauplii of each species vary somewhat in absolute length and breadth, and the length and breadth vary independently, giving slight differences in shape, but the number of cases noted is too small to be worth recording.

(C.) *The Carapace.*

The carapace is characteristic in most of the genera.

It is essentially an expansion of the dorsal side of the body in anterior, lateral, and posterior directions, the cuticle on the dorsal side of which is somewhat thicker than over the rest of the body.

It is shield-shaped and shallow in *Lepas* and *Conchoderma* (figs. 156, 157, 162), and has its anterior angles produced into two long and slender delicately striated horns (*frl.h.*) ; these are directed slightly downwards and forwards, and terminate in

may, perhaps, represent the invaginated caudal spine and adjoining folded cuticle just before a moult.—[13/7/93.]

an aperture bounded by slight anterior and posterior spine-shaped edges, between which are dorsal and ventral V-shaped gaps. The posterior margin of the carapace is produced into a long median caudal spine (*c.sp.*), articulated at the base and covered with slender spinelets.

In *Dichelaspis Darwinii* (fig. 168) the carapace is triangular; the caudal spine about the same length as in *Lepas* and *Conchoderma*, but the horns much shorter.

In *Balanus* the carapace apparently varies somewhat in shape. In *Balanus perforatus* (figs. 141, 142) it is rather broadly shield-shaped and distinctly convex, with a somewhat convex anterior edge, short fronto-lateral horns, and two short blunt teeth on its lateral margin just where it passes into the broad-based caudal spine; the latter is much shorter than in the foregoing genera, and is covered with minute spines.

In *Balanus improvisus* (22) the frontal horns are relatively about as long as in *Balanus perforatus*, but the carapace is apparently more triangular, and the caudal spine relatively shorter.

In *Balanus balanoides* the caudal spine and fronto-lateral horns are still shorter, and farthest removed from the Lepad type.

In *Chthamalus stellatus* (figs. 149-151) the carapace approaches a circular or ill-marked hexagonal form; the anterior and posterior margins are straight and parallel; the fronto-lateral horns as short as in *Balanus perforatus*; the lateral margins show two slight bulges at about two-thirds of the length of the carapace from the fronto-lateral horns; these mark the openings of two large lateral glands; the caudal spine has a very broad base with toothed margins, and is covered with rather short and strong secondary spines. In this species the thickening of the cuticle extends anteriorly on to the ventral side, and ends in an abrupt transverse line just in front of the frontal filaments.

In *Tetraclita porosa* [relying on FRITZ MÜLLER's figure (16)], the carapace is also nearly circular, and in general appearance much like that of *Chthamalus stellatus*.

In *Scalpellum vulgare*, according to the figures of DARWIN and LANG, the carapace is very broadly shield-shaped, with bulging sides and fairly short horns; the caudal spine is short in the figures, but it appears possible that this may be due to its being telescoped, especially as LANG describes the hairs on the appendages as telescoped.

In *Pollicipes polymerus*, as figured by NUSSBAUM, the carapace is very similar to that of *Scalpellum*, having prominent sides, fairly short horns, and a short caudal spine; how long the latter is does not appear from the description or figure, since it is not stated whether, in NUSSBAUM's Plate 2, fig. 3, it is protruded or not; probably it is retracted.

In *Verruca Strömia*, judging from SPENCE BATE's figures (though these are apparently somewhat diagrammatic), as that author has pointed out, the caudal spine is much longer than in *Balanus balanoides*. It is apparently about as long relatively as that of *Balanus perforatus*, which has the longest spine of all the Balanid species hitherto observed. The shape of the carapace closely approaches that of *Balanus*, and the fronto-lateral horns are apparently relatively about as long as in *Balanus perforatus* or *B. improvisus.*

It is thus evident that the carapace varies from a broad and convex shield-shaped or almost circular structure with short horns and caudal spine to a shallow more or less triangular structure with long horns and posterior spine.

Its surface is quite smooth in *Balanus perforatus*, but in *Conchoderma, Lepas,* and *Chthamalus* it shows a fine reticular structure, which I at first imagined was due to the outlines of minute ectoderm cells below it till I perceived that the ectoderm cells were too scattered to produce such marking. In cases where decomposition of the animal has set in the markings are lost, though delicate chitinous hairs remain. The markings are evidently not in the chitin, but belong, in some way not perfectly clear to me, to the more decomposable tissues immediately beneath. The meshes of the network are fairly large in *Chthamalus*, but very small in *Lepas* and *Conchoderma;* in the latter two forms its distribution gives rise to a tolerably definite pattern formed by a series of intersecting curves, the main lines of which run in definite systems, which have, in many cases, a marked relation to the points of attachment of fibres passing from the carapace to the walls of the alimentary canal.

The cuticle over the whole of the body is everywhere underlaid by the ectoderm, a layer of protoplasm (except where the muscles are attached), which has a finely granular appearance and is generally thin, but locally thickened in the neighbourhood of the large scattered nuclei, with their distinct nucleoli. At the sides of the carapace is a band pretty definitely marked off from the rest along which the ectoderm cells have a vesicular appearance.

At the base of the fronto-lateral horns are situated in all the species examined a pair of large glands—the fronto-lateral glands (*frl.gl.*)—which are probably to be regarded as specially developed ectoderm cells, though the embryological evidence on this point is not complete.

The gland has the same structure in all the species. It consists of two greatly developed cells which have strongly refractive chitinous. walls; these completely invest each cell except at the distal ends, where the chitinous sac ends in an irregular margin within the tubular horns. The two chitinous sacs which are respectively anterior and inferior, and posterior and superior, are closely applied together for the whole of their length, and together form a spindle-shaped organ extending from a point on the right or left side of the lower surface of the carapace above the sides of the stomach : the main and distal portion of each sac is either empty or filled with spherules of a transparent secretion, which are generally so numerous as to be pressed together into polyhedral bodies (*frl.gl.s.*). When found outside the horns these drops of secretion are spherical.

A considerably smaller portion of the proximal end of each sac is occupied by finely granular protoplasm, with a large nucleus (seen in *Lepas, Conchoderma,* and *Balanus*) with a reticular meshwork of deeply staining threads and particles (fig. 138).

The protoplasm, as it is traced away from the nucleus, becomes filled by droplets of the secretion, which finally merge together to form the larger spherules filling the sac. The posterior of the sacs was often found to be empty in *Lepas*.

The lumen of the horns is commonly partially interrupted in *Balanus* and *Chthamalus* (figs. 141, 142, 149) by structures looking like spines or hairs. It must have been this structure which led DARWIN to believe that the axis of the horn was occupied by a plumose process, regarded by him as corresponding to the prehensile antennules of the Cypris-stage. They are commonly attached only to one side, and are, I suspect, caused by the solidification of the peripheral part of the secreted spherules while still within the tube, and the bursting of these caused by the pressure of the secretion behind ; the partitions, at all events, closely resemble the walls of the spherules of secretion filling up the cavity of the glandular sacs.

The fronto-lateral glands are, accordingly, a pair of large unicellular and uninucleated glands.*

Posterior to these, in *Balanus perforatus* and *Chthamalus stellatus*, are a pair of similar but rather smaller glands (figs. 141, 149, *lat.gl.*), pretty definitely marked off from the rest of the tissues, but without the strong refractive investment of the fronto-lateral glands. These are attached to the carapace close to the point of insertion of the fronto-lateral glands ; they pass backwards and outwards to a point about one-third of the length of the margin of the carapace in front of the commencement of the caudal spine, and open externally by a pair of apertures very prominent in *Chthamalus*, and situated on a part of the margin bulging out beyond the rest (fig. 150).

These are the *lateral glands* : they are absent at this stage in *Lepas* and *Conchoderma*.

In all three forms a third structure occurs (*d.b.* in figs. 156, 157, 162), which, on account of its resemblance to the glands of the labrum to be described shortly, I am disposed to regard as glandular.

It consists of a short cord of very granular or vesicular tissue (*d.b.*), arising from the dorsal surface of the investment of the brain close behind the eye and passing upwards and forwards to the carapace. Its substance is distinct from that of the brain, and there seems no reason to infer any intimate connection between the two. In later stages it is clearly seen to be unicellular.

(D.) *The Labrum.*

The labrum (*lbr.*), though always large, varies considerably in relative size and form, and in details of structure, in the different genera. It is characteristic in *Lepas*, *Dichelaspis*, *Chthamalus*, and *Balanus*, and probably in other genera ; in *Conchoderma*, however, in agreement with the close affinity of this genus

---

* They appear to be essentially similar in structure to the two pairs of glands described and figured by GROSSEN in the tail region of the Copepod *Cetochilus*.

with *Lepas*, it is indistinguishable from the labrum of that species. In other genera it is unknown to me, and the published figures, with, perhaps, the exception of *Verruca*, are not sufficiently good to rely upon.

It is very large in *Lepas* and *Conchoderma*, where it extends to the posterior end of the body.

In *Dichelaspis* it is as long, but more slender and pointed. In *Balanus* it is relatively broad and short.

It consists in the above-mentioned genera of a proximal (*lbr.prox.*) and a distal lobe (*lbr.dist.*). Both are furnished with a row of long hairs on the upper surface (that turned towards the ventral side of the post-oral region) : these follow the margins for the greater part of the length of the labrum, but proximally turn inwards and converge towards the mouth.

The proximal lobe is ovate in shape in *Lepas*, *Conchoderma*, *Dichelaspis*, and *Chthamalus*, but broader and shorter in *Chthamalus* than in the others. In *Balanus* it is also broad and short, but its distal end is furnished with two small lateral lobes which give it a characteristic shape.

The distal lobe, though smaller than the proximal, is large and pentagonal in *Lepas* and *Conchoderma*, resembling in shape the section of a rather flat-topped haystack, and is provided with one large perforated median and two smaller lateral teeth on its expanded distal margin. In *Dichelaspis* it is narrow, parallel-sided, and pointed at the end. In *Chthamalus* it is large and parabolic, and supplied with two or three distinct median and two lateral teeth, together with three or four smaller ones between these on each side. In *Balanus* the median lobe is similar in shape though smaller, and supplied with two lateral teeth, one on each side.

In *Verruca*, relying on SPENCE BATE's figure, the basal lobe of the labrum is large and ovate ; the distal lobe small and pentagonal, and resembles *Lepas* and *Conchoderma* in the disposition of the teeth (one median and two lateral), but agrees more closely with *Dichelaspis* in the prominence of the median part bearing the unpaired tooth.

The labrum has sometimes been stated to be movable, but I never observed this, and satisfied myself that no muscles pass to it with the exception of short ones from the œsophagus.

The labrum is longer on its lower than on its upper side, the lower being attached immediately behind the eye, and the upper at about the level of the point of attachment of the antennæ.

The axis of the labrum is occupied by a peculiar gland (*ax.gl.*) confined to it, and opening at its distal end.

This occurs probably in the Nauplii of all the Thoracic Cirripedes ; I have observed it in *Lepas*, *Conchoderma*, *Dichelaspis*, *Chthamalus*, and *Balanus*. It is of different nature to the fronto-lateral and lateral glands. Though essentially similar in all these it varies somewhat in the different groups.

It is simplest in *Lepas anatifera*, *L. pectinata*, and *Conchoderma*, for which forms one description will suffice. It is readily visible in all specimens of these genera as a narrow cord of granular or vesicular matter running from the base of the labrum to its end. It is this, and not a simple groove as BALFOUR supposed, which has been taken by so many observers for the œsophagus.

The gland on careful examination is seen to be composite, and to consist of four elongated and probably unicellular glands, two of which, forming a pair, are attached to the sides of the base of the labrum some distance in front of the mouth (figs. 158, 159), and two lying close together, forming a central strand inserted immediately below and in front of the mouth; all three strands unite in the middle line, while still within the proximal lobe of the labrum, but, maintaining their independence, pass towards the end of the distal lobe of the labrum (within which, not far from their termination, they may become slightly expanded and diminish) to the central tooth by an aperture, on the summit of which they open to the exterior.

The contents vary somewhat, but generally consist of deeply-staining, coarse granules of uniform size, but in many cases a different appearance is caused by the presence of a number of vacuoles.

I observed in some cases one, and, as far as I could ascertain definitely, only one nucleus for each cell, generally placed near the proximal end.

In the remaining forms examined the contents of the glands consisted of finely granulated vacuolated material.

In *Dichelaspis Darwinii* (fig. 168), as far as I could make out from mounted specimens, the form of the gland approaches that of *Lepas* and *Conchoderma*, except that its distal end is expanded into a spindle-shaped body, filling up the greater part of the distal lobe of the labrum. One or more of the large vacuoles are generally present.

In *Chthamalus stellatus*, in which the labrum was only investigated in spirit specimens, I could only make out the distal portion of the gland; this is swollen (fig. 150) in a marked manner at the tip, as in the case of *Balanus*.

In *Balanus perforatus* (figs. 139*a*, 142, and 145) the gland consists of four pear-shaped cells which meet together at the distal end of the labrum, but in accordance with the diminution in length of the labrum the gland is shorter owing to the absence of the part corresponding to the proximal portion in *Lepas*. The nuclei are large and distinct, and furnished with a distinct intranuclear network, and generally resemble those of the fronto-lateral glands, but are smaller. The glands are here distinctly seen to be unicellular. A delicate axial fibre (figs. 136*d* and 145, *ax.gl,fi.*), apparently of the nature of connective tissue, runs from a small group of cells situated immediately in front of the mouth to the distal end of the axial gland, serving, perhaps, as a support for the gland cells.*

* These glands apparently correspond with those figured but not described by CLAUS (61) in the labrum of the Nauplius of *Branchipus*.

Immediately behind the labrum, as HOEK alone has stated for *Balanus balanoides*, and BALFOUR supposed in *Lepas fascicularis*, is the mouth, with a strong chitinous margin, which excavates with its lower half the base of the labrum. This side is also deepened by two grooves bounding a small median lobe between them, while the chitin is thinner and less conspicuous on the upper margin.

(E.) *The Setose Region.*

The setose region (figs. 144, 152, 160, and 163) presents certain characters common to the whole group, but the details, though constant in the species, are often distinctive of the genera.

Passing down the sides of the region from the mouth in an elegant lyre-shaped curve in *Lepas*, *Conchoderma*, *Chthamalus*, and *Balanus* is a band of long hairs (which may be double) bending in towards the middle line and immediately beneath the ventral insertions of the flexor muscles. These may be termed the flexor arcs (*fl.arc*).

In front of the flexor arcs, disconnected in *Chthamalus*, united together by two slight ventral loops in *Balanus*, and by a distinct band in *Lepas* and *Conchoderma* (where it is immediately followed by a second slight transverse band), a band extends on each side from about half way along the flexor arcs towards the ventral side of the region; these may be termed the *anterior arcs* (*ant.arc*). In the normal position of the parts they lie immediately above the labrum, forming with the hairs of that organ a sieve, preventing the escape of large particles from beneath it.

In *Balanus* and *Chthamalus* there is also a median group of hairs on the region above the base of the labrum and not far behind the mouth.

Behind the flexor arc are two short curved lateral bands, between the anterior ends of which are two slight curved bands in *Lepas*, *Conchoderma*, and *Balanus*, apparently represented by a single transverse band in *Chthamalus*. The lateral bands may be termed the *extra-maxillary arcs* (*ex.mx.arc*), and the smaller bands the *pre-maxillary bands* (*pmx.bd*.).

Posterior and nearly in a line with the former are two bands of long bristles passing obliquely upwards to unite on the dorsal side of the tail behind the anus : these are the *anal arcs* (*an.arc*) ; they belong properly to the thoracic region.

In *Balanus* the extra-maxillary bands are only slightly developed.

In *Lepas* and *Conchoderma* a *maxillary arc* runs from the point of junction of the extra-maxillary and anal arcs.

These terms are used to facilitate descriptions, and because some knowledge of the area will be required in discussing the morphology of the gnathites.

(F.) *The Tail* (*Thorax-abdomen*).

The tail (*ta.*) varies considerably in length and appearance in the different genera, and to some extent in the species.

In *Lepas anatifera, L. pectinata, L. fascicularis* (28), probably in *L. anserifera* (5), and in *Dichelaspis*, it is long and simple, and covered with fairly long secondary spines.

In all the examples of *Conchoderma* investigated it was similar, but bifurcated at the tip.

In *Verruca* (21) it is apparently similar to *Conchoderma*, but shorter.

In *Chthamalus stellatus* it is short and the spines few.

In *Balanus* the secondary spines are much shorter. In *Balanus perforatus* the tail is rather short, in *B. improvisus* (22) it is still shorter, and in *B. balanoides* (30) apparently very short.

At some distance behind the base of the tail in all the species are a pair of very strong ventrally and laterally directed spines, with toothed postero-ventral and antero-dorsal edges, the teeth being much more numerous and marked on the ventral side in most species. The teeth of the posterior margins are continuous across the middle line.

These spines mark the commencement of the abdomen, and, as will be seen later, correspond partly to the caudal appendages of the Cypris stage. The posterior end of the head region is marked off by the maxillary arc.

The thorax occurring between these limits is marked laterally by the anal arc, and is generally provided with fewer and shorter spines than the caudal or abdominal region.

(G.) *The Appendages.*

Fortunately the appendages, which might be supposed to need the longest descrip-tion in a comparative account, show no important difference in any of the species I have examined.

It is not a little remarkable that with the exception of the presence of some short simple hairs in *Lepas*, which are absent in *Balanus*, and of slight differences in the relative proportions of the parts, the structure of the appendages is exactly the same in each of the species. There are the same number of joints on corresponding branches or basal pieces, and the same number of bristles, spines, or teeth on each of the joints; these processes are numerous, and present great variety in structure, but the differences are repeated in each species, so that there are precisely the same number of simple bristles, plumose bristles, simple spines, plumose spines, teeth, &c., in the Nauplius of every species at this stage.

This is true of such different forms as *Lepas anatifera, L. pectinata, Conchoderma virgata, Chthamalus stellatus, Balanus perforatus,* and probably all the rest of the Thoracica.

Few who have studied Cirripedian development have attempted to minutely characterize the appendages, and a careful comparison of the species has given results, which, while they agree perfectly together, only partially agree with the statements of those who have. This, however, is not difficult to understand considering the difficulty of counting numerous hairs on such minute forms, and I must say that it was only a perception of the very close agreement between all the species that saved me, as I believe, from error myself. For I had only to see a difference in two of my sketches to discover upon re-examination of the object that a mistake had been made in one or other.

In *Balanus improvisus*, the characters of the appendages, as given by BUCHHOLZ, much resemble that given below; in *Balanus balanoides* (30) the description agrees even more closely, while in LANG's description of the Nauplius of *Balanus perforatus*, which is the only account in which minute details are given, the agreement is almost perfect, but, still not quite exact, but as this is one of the species I examined this discordance must disappear.

In describing the appendages it will be convenient to recognize the following—*hairs*, small and simple; *bristles*, much stronger; *plumose bristles*, with hairs on one or both sides; *spines* and *plumose spines* much stiffer; *conical processes* and *teeth* still stronger; and the *gnathobase* as a movable biting piece.

It will abbreviate description to state that all the bristles, spines, &c., at this stage, arise on the ventral sides of the appendages, with the exception of the line of simple hairs on the dorsal sides of the second and third pairs of appendages; and that the bristles and spines curve inwards towards the centre of the ventral surface. It is also to be noted that the secondary hairs on the bristles are given off in the *dorso-ventral* plane on one or both sides.

The antennules or first pair of Nauplius appendages are a pair of uniramous, elongated appendages, generally directed downwards and forwards, parallel to the long axis of the body, but during motion moved to a position at right angles to this.

They consist of four joints, of which the first or basal is of truncated conical shape and devoid of hairs, the narrow end fitting by means of a flexible membrane into a circular opening in the body wall situated opposite the extreme anterior end of the labrum, in the angle between that organ and the lobes of the brain; they are thus distinctly pre-oral in position; the remaining joints are cylindrical; the second bears a simple bristle at the distal end; the third is much the largest, and gives off about its middle a plumose bristle, and at its distal end two bristles, one simple, one plumose; the last joint terminates abruptly; at its distal end is a dorso-ventral row of four bristles, the second ventral one plumose, and rather stronger than the rest, which are simple. In *Balanus* there are at this stage no dorsal bristles or hairs, but in *Lepas* there is a bunch of hairs at the distal end of the third joint.

The antennæ are large and consist of a two-jointed *protopodite*, an *endopodite*, and an *exopodite*.

The protopodite is two-jointed and very stout. Its proximal joint is attached to the body, not far from the anterior corners of the mouth, by a rather narrow neck, which soon expands into an almost spherical distal portion; this is produced on its ventral and inner side into a very strong conical process or gnathobase (*gn.*), with two sharp teeth at its apex, surrounded with small spines and hairs. This process works with the limb in a horizontal circle towards or away from the mouth, close behind which the teeth are situated when the appendages are directed backwards; it has also an independent motion of its own, being articulated with the globular head of the joint, and moved, as I have been able to observe, backwards and forwards by means of a powerful muscle (fig. 142, *gn.m.*), for the reception of which the joint is swollen. The second joint is also rounded but much smaller than the first, and gives rise on its inner and ventral side to a powerful hairy spine, with a strong conical base, which is furnished with two short bristles, one simple, the other plumose. The dorsal edge of both joints has in *Lepas* a line of short hairs, scarcely represented in *Balanus.*

The endopodite (in accordance, perhaps, with the function of the spines of its base) is not sharply separated off from the protopodite, the basal joint not being distinctly articulated. It consists of three joints; the basal one is rounded, and has on a prominence on its inner and ventral side two long spines with numerous hairs arranged regularly down both sides. The second joint is smaller and cylindrical, and has two ventral stiff plumose bristles placed side by side at its distal end. The third joint is oval, and provided with three bristles, the most ventrally situated simple, and the two others plumose.

The exopodite is longer than the endopodite, and lies dorsally to it. It consists of nine joints, of which the first seven are cylindrical, and diminish regularly in diameter from the proximal extremity; the eighth is conical, and the ninth minute. The dorsal side of the whole exopodite is marked by a band of short hairs continuous with those on the protopodite. The first joint has no other appendages; the second and third have a ventral row of hairs; the third has, in addition, a simple bristle at its distal end; the next four joints (4 to 7) have each at their distal end a long bristle, plumose on both sides. The bristle of the eighth is very strong and almost as wide at its base as the joint itself, and is plumose on both sides in *Balanus,* and more strongly so on one side in *Lepas;* the bristle of the ninth is smaller and simple. The axis of the exopodite is generally curved somewhat dorsally.

If the term mandible were used in a descriptive sense, the second pair of appendages of the Cirripede Nauplius should certainly be so termed, but since the third pair probably morphologically represents the first and strongest pair of jaws in the adult, the name may be reserved for the latter.

The *mandibles*, or third pair of appendages, are smaller than the antennæ, and consist, likewise, of a protopodite, endopodite, and exopodite.

The protopodite is two-jointed. The basal joint is attached by a rather narrow neck close behind the antennæ, and behind and outside the mouth; it enlarges distally to form a rounded head which gives off on its ventral and inner side a strong spine, plumose on two sides, and which works horizontally (with the limb only) behind the mouth, and below the bidentate process of the antennæ. The second joint is large, flattened from side to side, and expanded distally ; on its lower side is a semicircle of hairs (double in *Lepas*) with the concavity turned inwards ; this joint also has two spines, which are bent inwards, and plumose on two sides, the proximal one being stouter, especially in *Balanus ;* in *Lepas* both rather resemble stiff bent plumose bristles.   .

The endopodite is two-jointed ; and, as in the case of the antennæ, not so distinctly articulated as the somewhat shorter and more dorsally situated exopodite, and not sharply marked off from the protopodite. The first joint is short and expanded ventrally ; it bears on its lower side a semicircle of hairs in *Lepas*, in addition to which is a distal group of hairs ; the prominence bears three plumose spines bent inwards, one of which, situated most dorsally, is stouter than the other two, especially in *Balanus*. The distal joint is bluntly conical, with a ventral notch not far from the end ; this gives insertion to two stiff bent bristles placed side by side, one on the outer, and one on the inner side, the former plumose, and the latter simple ; at the base of this joint is a small semicircle of hairs on the outer side in *Lepas ;* at the end of the joint are three stiff bristles, two more ventrally situated, placed side by side, and one terminal ; all three are simple.

The exopodite is five-jointed and longer than the endopodite. The joints taper off distally, each bearing a bristle ; that on the first joint is simple and shorter than the rest ; the next three are plumose on both sides, and the fifth is the direct continuation of the small spindle-shaped fifth joint, and has long secondary hairs on the ventral side only. The dorsal side of the exopodite has also a row of simple hairs.   .

An axial portion can be distinguished within each hair or process, which has been regarded as a nerve ; it is, however, nothing more than the hair of the next stage, as can readily be seen just before a moult.

TABLE showing Number of Joints and Processes in the Appendages of the Cirripede Nauplius of Stages 1 and 2.
(t = terminal; l = lateral.)

Antennules.

	Stage 1.	Stage 2.
Joint 1 .	...	...
„ 2 .	1 bristle (t) .	1 bristle (t) .
„ 3 .	2 bristles (t)	2 bristles (t)
„ 3 .	1 bristle (l) .	1 bristle (t) .
„ 4 .	4 bristles (t)	4 bristles (t)
„ 4 .	...	...
Total .	8	8

Antennæ.

	Stage 1.	Stage 2.
Protopodite joint 1	1 process .	1 process .
„ „ 2	1 spine .	1 spine .
„ „ 2	1 bristle.	2 bristles
Endopodite „ 1	2 bristles	„
„ „ 2	„	„
„ „ 3	„	„
Exopodite „ 1	...	...
„ „ 2	...	...
„ „ 3	1 bristle.	1 bristle.
„ „ 4	1	1
„ „ 5	1	1
„ „ 6	1	1
„ „ 7	1	1
„ „ 8	0	1
„ „ 9	0	1
Total	15	18

Mandibles.

	Stage 1.	Stage 2.
Protopodite joint 1	1 process .	1 process .
„ „ 2	1 spine .	1 spine .
„ „ 2	1 bristle .	1 bristle .
Endopodite „ 1	1 spine .	1 spine .
„ „ 1	1 bristle .	1 bristle .
„ „ 2	2 bristles (l)	2 bristles (l)
„ „ 2	3 „ 3	3 „ 3
Exopodite „ 1	1 bristle .	1 bristle .
„ „ 2	1	1
„ „ 3	„	1
„ „ 4	„	1
„ „ 5	0	1
Total	14	15

(H.) *The Alimentary Canal.*

The *alimentary canal* already possesses the three divisions characteristic of that of the adult. These have been noted by MÜNTER and BUCHHOLZ. They are the *œsophagus (stomodæum)*, the *stomach* (mainly *mesenteron*), and the *intestine (proctodæum)*. It is not possible, I think, to distinguish, as these authors have done, a posterior section or rectum.

The *œsophagus (œs.)* commences with a mouth provided with a thickened chitinous margin, and runs forward, as DARWIN has observed, from immediately behind the base of the labrum, as a bent tube, the course of which is indicated in the figures 145 and 158. The first part is horizontal, and leads close to the lower sloping surface of the head, which may be described either as the base of the labrum, or as a part of the head continuous with it. It then bends round in a gentle curve, and passing between the circum-œsophageal connectives into a short vertical section placed immediately behind the brain, which can be nearly always seen in optical section in both dorsal and ventral views (fig. 157). Its termination projects (as in the adult) backwards into the stomach (figs. 136c and 139a). The tube is slender and consists of a single layer of polyhedral cells (figs. 136c–137d), each with a relatively large nucleus, often containing a single nucleolus; about four to eight of these cells are seen in transverse section; the lumen is distinct, and circular or quadrangular in section. When living the œsophagus is seen to be transversely striated, and to be undergoing peristaltic contractions (which have led to its being regarded as a heart). The appearance and movement are due to the presence of a single layer of circular muscles, which are readily seen when the motion ceases, or in stained specimens. They consist of a number of simple spindle-shaped fibres, apparently unstriated at this stage, short and broad when contracted, and much elongated at other times. The · œsophagus is attached to the walls of the labrum by a number of contractile fibres (figs. 145 and 158) which radiate out from, and serve to dilate it.

The *stomach (st.)* consists of a large simple globular or spheroidal sac, usually of a green or brownish colour, and occupying a considerable part of the anterior of the body of the Nauplius.

It has a distinct enteric cavity which arose for the first time just before the hatching of the Nauplius. It has now also an ordinary cellular structure, and as may readily be determined the cells arise by direct transformation of the yolk-endoderm pyramids. Just before hatching it will be remembered that the whole stomach consisted of a number of yolk-endoderm cells, each consisting mainly of yolk granules but containing a single nucleus. Each pyramid has now become directly transformed into an endoderm cell (fig. 141) by conversion of the yolk granules into protoplasm. The endoderm cells at first large (fig. 141), soon become much smaller by rapid division (figs. 156, 136a–d), and all traces of the yolk disappear. The endoderm is thus not, as has been supposed by HOEK, a new formation independent of the yolk segments. My account agrees rather with that of NASONOV, who describes the nuclei in the endoderm segments as becoming peripheral, the granules diminishing in number, and the endoderm segments becoming smaller as a cavity arises between them.

The wall of the stomach consists of a single layer of cells of two sorts : the first are polyhedral cells larger than those of the œsophagus and intestine, and containing a large round nucleus with a nucleolus (figs. 136a-d) ; they occupy the whole wall of the stomach, except a small posterior section, and are directly derived from the yolk-endoderm cells. In the neighbourhood of the opening into the intestine the cells forming the posterior wall of the stomach ($st.gl.c.$) are much narrower and higher, forming a columnar epithelium, the nuclei of which are oval and arranged radially ; the protoplasm of these cells is distinctly radially striated and stains much more deeply than the rest of the cells of the stomach (figs. 136a, 136b) ; these cells are part of the proctodæum. The opening of the pyloric end of the stomach is small (fig. 136a) and situated dorsally, owing, apparently, to a downward growth of the hinder wall of the stomach. The ventral side of the pyloric end projects below the anterior end of the intestine (figs. 136a, 136b). The statement of BUCHHOLZ that the stomach is furnished with circular muscles requires limitation, as they occur only on the portion at the posterior end. These muscles are quite similar to those of the œsophagus and intestine ; circular muscles are thus apparently limited to the part of the gut arising from the stomodæum and proctodæum. Extending from the posterior end of the ventral prolongation of the pyloric end of the stomach to the middle of the intestine is a group of simple unstriated longitudinal muscular cells (figs. 136a, 136b, 142, 146).

The *intestine* (*int.*) varies in appearance from a narrow, tubular, distinctly trans-versely-ringed organ to a broadish oval sac, according to the state of contraction of the muscles surrounding it. It consists, like the œsophagus, of an epithelial and muscular layer. The former is a simple layer of low, rounded or flattened, cells (figs. 136a, 136b), not always distinctly marked off from one another ; the nuclei are small and contain a single nucleolus. The muscles are simple, unstriated, spindle-shaped fibres with a single nucleus, varying much in appearance according to the state of contraction. The anus, as already stated, is a narrow aperture, placed between the tail and caudal spine.

In addition to these epithelial and muscular elements, the walls of the alimentary canal are completed by mesodermic cells, which will be referred to again.

The alimentary canal is attached between its extremities to the body walls by fibres, some of which are probably contractile ; just before the anus is a transverse sheet of tissue, attached to the body walls all round the anus (figs. 141, 156, *an.dil.*), and probably including many contractile fibres.

(J.) *The Nervous System.*

The *nervous system* at this stage differs somewhat in the Lepads (*Lepas anatifera*,

L. pectinata, Conchoderma virgata, Dichelaspis Darwinii) and Balanids (*Balanus perforatus, Chthamalus stellatus*), but the structure is very uniform in the members of the former family.

None of the existing descriptions of the nervous system are complete; most observers have seen the brain alone, and have described and figured even that incorrectly. KROHN and NUSSBAUM alone appear to have seen the circum-œsophageal ring and sub-œsophageal ganglion, the former in a Balanid Nauplius, and the latter in *Pollicipes;* but the figures of both are diagrammatic, and scarcely any details are given.

In order to make out the nervous system satisfactorily, it is necessary to stain and clarify the more transparent Nauplii, and to cut sections of the more opaque.

In the Lepads (figs. 154, 156–159) it consists of a small brain of two lobes (*br.*), closely applied in the middle line, and staining more deeply than the circum-œsophageal connectives arising from them, owing to the more numerous ganglion cells present. Each lobe is spherical, and situated close to the ventral surface ; it passes backwards into one of the two connectives (*c.o.c.*). The latter have ganglion cells all along their course, but present in much smaller numbers than in the ganglia. The connectives diverge and form an elongated loop on each side of the œsophagus and terminate immediately behind the mouth in two triangular* ganglia (figs. 159, 160), closely applied in the middle line to form the sub-œsophageal ganglion (*sub-œs.g.*), the posterior wings of which project (in the direction of connectives to be formed later) close beneath the flexor arcs about as far back as the level of the posterior end of the stomach, but not so far as the end of the labrum ; the sub-œsophageal ganglion is thus concealed both in dorsal and ventral views.

Each lobe of the brain is excavated by the spherical base of the frontal filaments, and supports on its dorsal surface close to the middle line one-half of the Nauplius eye.

All parts of the nervous system at this period are in close connection with, or rather form thickenings of, the ectoderm.

In *Chthamalus* (fig. 149) the nervous system resembles that of Lepads, with the exception of the addition of the pair of accessory lobes (*br.acc.l.*) described in *Balanus.*

In *Balanus perforatus* (figs. 137c to 139a, 141, 143, 145) the nervous system is more complex ; it here consists of a complex brain, circum-œsophageal connectives, and a sub-œsophageal ganglion, all in the closest relation with the ectoderm.

The brain consists of a number of lobes, of which two—the *anterior lobes* (*br.*)— resemble and correspond to the simple lobes of Lepads. These are more or less reniform, and closely applied to one another anteriorly, but separated behind by two other lobes. They are excavated anteriorly by the bases of the frontal filaments, and support close to the middle line the pigmented plates of the Nauplius eye ; the

---

* The shape was not made out in *Dichelaspis.*

lateral portions, not as yet distinct from the rest, will form the compound eyes, and may be termed the *optic tracts.* The anterior lobes are thus sensory in function.

Between the hinder ends of these lobes, and distinct from them, are two smaller hemispherical lobes (figs. 137*d*, 141, 143, *br.acc.l.*), closely applied to one another in the middle line, and which may be termed the *accessory lobes.* I have failed to recognize their significance.

The whole of the anterior lobes consist of small ganglion cells, the nuclei of which are closely crowded and show a number of deeply staining granules. The same may be said of the accessory lobes ; the latter, however, rest upon a large spherical mass of grey matter (figs. 137*a* to 137*c*, *br.c.l.*), also associated anteriorly with the anterior lobes, and posteriorly almost in connection with the grey matter of the connectives.

On the ventral side of this *central lobe* is a second accumulation of ganglionic matter on each side (fig. 137*a*, *br.p.l.*), composed mainly of ganglion cells and forming the posterior section of the ventral portion of the brain ; these may be termed the *posterior lobes.* They meet in the middle line and give off posteriorly the two circum-œsophageal connectives (figs. 143, 145, 137*a*, *c.o.c.*). The circum-œsophageal connectives separate and pass in an oval loop round the œsophagus ; they consist mainly of grey matter, but have a group of ganglion cells all along the ventral and outer border (figs. 136*a*, 137*b*).

The connectives give off on their ventral side shortly after leaving the brain two ventral lobes—the *nerves to the labrum* (*lbr.n.*)—which pass into (figs. 136*c*, 136*d*) and occupy the dorsal section of the sides of the proximal lobe of the labrum ; they run nearly to the end of this lobe (fig. 145). These branches consist mainly of ganglion cells ; their origin and fate are closely connected with those of structures apparently of glandular nature, more fully developed at a later date.

The connectives have in front few ganglion cells, but these soon increase in number, and limit the grey matter to their inner and dorsal side (figs. 137*b*, 137*c*) ; the connectives unite behind the mouth and form a broad and thick plate of ganglion cells, bilobed and still containing grey matter anteriorly (fig. 137*b*), but simple posteriorly (figs. 137*c*, 137*d*) ; the posterior section finally becomes thinner and disappears in the setose area.

The antennules arise precisely at the level of the posterior lobes, and are in close connection with them, a short prominence of the lobe (fig. 137*a*) sometimes projecting into the cavity of the appendage.

The antennæ arise at the level of the circum-œsophageal connectives (fig. 137*a*), and the ganglion cells on their outer and lower border sometimes project slightly into the cavity of the appendage.

The mandibles arise just opposite the anterior end of the large sub-œsophageal ganglion with which they are in close relation (fig. 137*d*).

(K.) *The Sense Organs.*

Of sense organs two kinds only can be detected at this stage. These are the Nauplius eye and the frontal filaments.

The *eye* (*Npl.eye*) consists in all the species of two oblong black pigment plates generally placed together (but sometimes separated by a small interval) in the middle line, like the two halves of an open book, each on the dorsal side of one of the anterior lobes of the brain, and not separated from the latter by an interval as figured by NUSSBAUM. Sometimes, especially when the eye is injured, a reddish colouring material may also be seen. I could detect no special cells in connection with these plates ; the ganglion cells below are apparently similar to those elsewhere.* The lens stated by a number of observers to exist, but denied in *Balanus balanoides* by HOEK, is equally absent in all the species I examined.

The (*fr.fil.*) are two delicate transparent filiform organs inserted, as FRITZ MÜLLER (16) and HOEK (30) have supposed, directly on the brain.

They consist of a cuticle,† separated from that of the rest of the body by flexible membrane, and of an axial strand. The sheath or cuticular portion consists of two joints, and is never multi-articulate as figured by SPENCE BATE ; the basal joint is short and conical, and the distal filiform and closed at the end. They are directed downwards and a little forwards, the distal joint having a ventral flexure: The basal portion arises on the ventral surface of the body at the level of the eye, and is situated on the ventral surface of the anterior lobe of the brain, inside which its continuation expands in the form of a spherical vesicle seen on each side of the eye (figs. 137b, 137c, 141, 142, 145, 149). In most cases the contents of the sheath are not distinctly seen, but in favourable sections a finely or coarsely granular portion is seen occupying the axis of the filament, and expanding within the spherical base. This axial portion much resembles in appearance the grey matter of the brain, and is probably of a nervous nature and in communication with the brain at the base of the sac, though I was not able to observe this connection.

There are no muscles passing to the frontal filaments, which can therefore be moved only in a passive manner.

(L.) *The Muscles.*

The muscles proper to the wall of the alimentary canal have already been described. There are others which are equally unstriated and which cause dilatation of different parts of the gut. These are most prominent in connection with the œsophagus, which they connect with the wall of the labrum (figs. 145). They are also found in connection with the hind-gut, and nucleated fibres, probably contractile, are found in the transverse membrane close to the anus, which they serve to dilate.

* Special cells in connection with the pigment plates are seen in some of my best preparations of later Nauplius stages.

† The cuticle, like that of the rest of the body, is insoluble in hot strong caustic potash.

The disposition of the striated muscles has been described in most detail by MÜNTER and BUCHHOLZ, and by HOEK, and the accounts of these authors agree in most points with my own. The muscles, however, stated to go to the fronto-lateral horns, do not exist, those of the antennules, or the glands of the horns, having been apparently mistaken for such. No muscles pass to the labrum, and any movements that organ undergoes must be brought about accidentally. It seems probable, however, that the distal lobe can move on the proximal, since two strong fibres (figs. 142, 145) (possibly unstriated muscles) pass from the dorsal sides of the proximal lobe of the labrum to the base of the distal in *Balanus*. As stated by the above mentioned observers, muscles pass from the median dorsal line of the carapace to the limbs (*app.d.m.*). They radiate out from the central area of the carapace as six systems, one passing through the dorsal section of the body cavity to each limb. The muscles to each limb do not all arise close together; thus, for instance, some muscles going to the mandibles arise in front of some going to the antennæ. There is on the ventral side an almost equally strong system of muscles (figs. 137b, 137c, 154, *app.v.m.*) which have been overlooked owing to their being hidden by the stomach in dorsal views and by the labrum in ventral views. They are readily seen when ventral views are obtained of Nauplii in which the labrum has been bent forward, and in lateral views, or even in dorsal views, by focussing deep down in transparent specimens. The figures will sufficiently indicate the disposition of these twelve groups.

The powerful flexor muscles (*fl.ta.*) (in ventral and lateral views) of the tail pass from the carapace close to the attachment of the secreting end of the fronto-lateral glands downwards and backwards to the base of the tail. They have been already correctly described by BUCHHOLZ and by HOEK. The muscles to the limbs are all transversely striated (fig. 138a). They are inserted directly on the cuticle at both ends. The histology of the several species I observed much resembles that described by HOEK in *Balanus balanoides*. The transverse striations are clear, and each fibre has one or more, commonly two, protoplasmic corpuscles on one side.

Of a number of muscles limited to the limbs and causing movement of one part on another, one only needs special mention. This occurs in the second joint of the pro-topodite of the antennæ, and is a short powerful fan-shaped muscle causing motion of the strong gnathobase of this appendage (fig. 142).

(M.) *Other Mesodermal Elements.*

The body cavity is bounded on one hand mainly by the ectoderm of the body walls and appendages, and on the other by the walls of the gut. At intervals more or less flattened mesodermic cells are scattered along these walls as well as on the brain, gland-walls, and muscles (figs. 149, 156). Others stretch across the cavity from one wall to the other, or from one organ to another (figs. 149, 151, 156); these occur throughout the body, but are especially distinct in the anterior part; here bi-, tri-, or multi-polar cells lie suspended by fibres between the structures mentioned. Though often single, mesodermal cells are very commonly distributed in groups ; one group is generally found in the angle between the fronto-lateral glands and the front wall (fig. 156). Four connective tissue bands are very conspicuous and constant in position in *Lepas* and *Conchoderma;* they run from the brain to the dorsal surface of the carapace near the front margin, as shown in all the dorsal views of these species ; the two outer are specially prominent, and contain three or four cells about the middle of their length ; they branch at their ends into a star-like series of fibres (*st.b.*). Two more of these stellate bodies are also found near the posterior end of the line bounding the lateral zone of the carapace.

In certain regions there is a peculiar tissue (*ves.t.*) which is difficult to understand ; this consists of an excessively delicate, vesicular, transparent, and colourless tissue, prominent at the sides of the intestine and stomach (figs. 141, 156), and occurring also in the caudal spine, tail, and in front of the fronto-lateral glands. It consists of very thin walls, cubical, or polyhedral vesicles, only staining slightly with reagents. In the tail and caudal spine a small quantity of granular staining matter, with one or two nuclei, can be detected in association with it ; elsewhere the tissue is not sufficiently distinct from the neighbouring tissues for examination ; it is probably to be regarded as a vesicular connective tissue.

The nuclei of the mesoderm cells are generally fairly large, the cells themselves not present in great numbers, and their boundaries tolerably distinct.

In the labrum there are, in *Balanus,* a number of cells (figs. 142, 145), into close connection with which the nerve of the labrum comes ; the nuclei of these appear very similar to those of the nervous system, but generally stain a little less deeply ; they occur all along the sides of the proximal lobe, and form a band extending transversely across its distal end.

In the Lepads (*Lepas, Conchoderma, Dichelaspis*) they form a small accumulation of cells (figs. 157, 159), (*lbr.c.*) embracing the diverging arms of the axial gland.

Of cellular elements floating about in the plasma at this stage I have no observations, though they occur sparingly in later stages.

Oil globules are not unfrequently seen embedded in the outer wall of the stomach.

(N.) *The Ectodermal Thickening of the Tail, or " Ventral Plate."*

The ectodermal thickening of the tail (*ta.th.*) has been passed over in silence by most authors. HOEK (30) alone has seen and figured it (in *Balanus balanoides*), but gave no interpretation of it. GROBBEN (35) has given a different figure, and states that HOEK's description is inexact ; he describes and figures a small band of mesoderm cells placed on each side below an ectodermal thickening, and consisting in the Nauplius examined of three cells, the most posterior of which was largest. He supposes from this that the mesoderm of *Balanus* arises from two pole cells. Now it follows from the foregoing description of the origin of the mesoblast at an early stage that this is not so, and I believe, moreover, that GROBBEN's figure is much less correct than the one of HOEK he criticises.* I tried the method recommended by GROBBEN for small Entomostraca (BEALE's carmine), but found the maceration too great in the case of Cirripede Nauplii, while specimens either unstained or lightly stained with borax carmine gave much better results.

The ectodermal thickening in *Balanus perforatus*, at Stage 1, is shown in fig. 146, and at a later period (Stage 2) in fig. 147. It consists at first of two plates of considerably-enlarged ectoderm cells which extend in a single layer from shortly behind the flexor muscles to the end of the tail. They are distinct in front, being separated clearly by a median groove, but unite together as the tail narrows behind. In Stage 1, the cells are relatively larger and fewer, they possess very distinct, clear, and large nuclei, each generally with a large nucleolus, and are arranged more or less in longitudinal (about four visible ventrally) and transverse rows. The anterior margin is nearly transverse. Well on in Stage 2, the cells are much more numerous and smaller, but still show an arrangement in longitudinal rows. In both stages the thickening extends obliquely upwards and forwards, but does not pass above the level of the intestine, so that the bands are truly ventral. The arrangement of the cells in longitudinal rows appears to indicate transverse division of a primitively smaller number of cells arranged transversely.

In *Lepas* and *Conchoderma* the thickening is of much less extent. Over the greater part of the tail the ectoderm has its usual characters, but in the anterior region just beneath, and closely following the extra-maxillary bands, are two oblique rows of thickened ectodermic cells forming the rudiments of the ectoderm and mesoderm probably of the fourth pair of appendages alone (figs. 160, 163, &c.).

In *Dichelaspis* and *Chthamalus* the thickening is similarly represented by a minute group of cells.

The appearance of these bands in *Balanus* strongly reminds one of the mesoblastic bands, so well-known to embryologists in other types. They lie, however, as a simple layer of cells (figs. 136b, 139c) immediately beneath the cuticle, and pass in front into ordinary ectoderm cells (fig. 139b). They are rather to be compared with the greatly-enlarged and symmetrically-disposed cells described by DELAGE in the same region in *Sacculina*, and seen by myself in *Peltogaster*. The fate of these cells will be traced in the second part of this work.

(O.) *The Body Cavity.*

The body cavity at this stage is a space (*b.c.*), devoid of a special epithelial lining and filled with a plasma containing few or no corpuscles. This is situated between the body wall and the alimentary canal, and extends into the labrum, appendages, tail, and caudal spine.

* It seems to me probable that the Nauplius examined by GROBBEN was that of *Chthamalus stellatus* and not a *Balanus* at all, in which case the difference between his and HOEK's description is more easily understood.

PART IV.—PHYSIOLOGY.

The physiology of the Nauplii is practically the same for all stages, and the following remarks apply equally to the earlier or later stages.

(A.) MOTION.

The movements of the Nauplius are brought about exclusively by the activity of the three pairs of appendages.   The three pairs are all moved together in a horizontal direction ; the bristles and hairs on the appendages are vertically arranged so as to give the greatest effect to this stroke.   The Nauplius thus moves by a series of jerks which follow one another with great rapidity in the newly-hatched Nauplii, but more slowly afterwards, especially in the loosely-built Nauplii of the Lepads, where, in Stage 2, the movement is comparatively slow.   In *Balanus perforatus*, the Nauplius, after the first moult, moves at the rate of about one millimetre per second, a rate equal to rather more than one hundred times it own length per minute, the strokes being at the rate of several per second.

The tail and caudal spine act as a rudder ; when in a line with the body they serve to steady and straighten the course, which is often very direct.   When the motion is to be suddenly arrested and reversed, the tail and caudal spine are sharply bent down at an angle by means of the flexor muscle, and the Nauplius turns a somersault.

(B.) NUTRITION.

The small and chitinized mouth is situated, as has been seen, at the base of the labrum.

I have no direct observations on the nature of the food taken in the natural state, nor do sections of the stomach show anything recognizable.

The fact that the stomach is often green has led to the statement that the Nauplii feed on plants ; it appears to me possible that the green colour is simply due to the alteration in colour of the original yolk.

The small size of the mouth indicates that only small bodies or substances in solution can be taken up.   The presence of powerful jaws on the appendages indicates that bodies of a resistent nature have sometimes to be masticated.   The abundance of stiff spines and bristles on the inner side of the endopodites, their position at some distance behind the mouth where they would not be of much use for mastication, and the curvature towards the space behind the mouth, seem to suggest the function of holding bodies of some size.   DARWIN has already expressed his belief that these spines are adapted for grasping rather than masticating.   The rich plumose character of the hairs on the inner sides of the appendages, and the direction of the hairs on the setose region and on the labrum, appear to indicate a function of retaining small particles within the sphere of attraction of the œsophagus. The œsophagus, as previously mentioned, undergoes continuous contraction and dilatation by means of the muscles with which it is furnished, and constitutes a suction-pump which causes an intermittent stream of water to enter the mouth.

It seems probable that the mode of nutrition is as follows :—The powerful strokes of the three pairs of appendages sweep backwards and inwards any small organisms or particles entangled between the network of hairs and bristles ; the œsophageal pump must cause a current towards the mouth, and the anal and other lateral rows of hairs will prevent the bodies from passing dorsally ; the presence of the forwardly-directed hairs of most of the setose region will hinder their escape backwards. Any substances in solution or small particles in suspension will be thus drawn into the mouth, while the larger bodies or organisms held fast by the spines of the endopodites, and possibly paralyzed by the secretion of the axial gland of the labrum, will be torn up and masticated by the powerful movable gnathobases of the antennæ, and the fragments, retained by the hairs of the labrum and setose region, sucked into the mouth.

The Nauplii are apparently predaceous, and I imagine their food consists, in addition to the more minute microscopic organisms, of small soft-bodied or tolerably resistent animals, such as would occur in the pelagic waters inhabited by the Nauplii.

That minute organic particles and substances in solution and suspension are taken into the body I have experimental evidence to show.

Though the Nauplii of *Balanus*, *Chthamalus*, and *Lepas* refused to eat starch, they greedily took up water containing carmine, indigo-carmine, methyl-blue, litmus, &c. In the case of the first three substances considerable accumulations were formed after a short time in the stomach and intestine, but the greater part of the carmine was passed out unabsorbed.

Litmus solution either remained blue or turned faint red in the stomach and hind gut, this showing the presence at times of an acid digestive juice ; the secretion of this probably takes place from the deeply-staining striated columnar cells forming the hind wall of the stomach.

The fæcal matter is expelled by the contractions of the pyloric portions of the stomach and of the intestine.

As in other Nauplii there were no special organs of circulation, the movements of the œsophagus and gut serving, no doubt, to fulfil this function, as well as that of respiration.

(C.) SECRETION.

In addition to the glandular epithelium of the stomach the only important secreting cells are the unicellular glands opening on the carapace and labrum, viz., the fronto-lateral gland, the lateral gland, and the axial gland of the labrum. It is only after the first moult that the fronto-lateral, and probably the other glands open to the exterior, though they are fully developed and contain abundance of the secreted globules while the Nauplius is still within the egg membrane.   The secretion of the fronto-lateral and lateral glands consists of transparent globules provided with a resistent pellicle, and which, though closely pressed against one another into polyhedral bodies within the cavity of the gland do not fuse ; they are not dissolved by water, alcohol, weak acids or alkalies ; they show no acid or alkaline reaction, and take up no colouring matter.   The membrane of the gland itself and of the globules is possibly of a chitinous nature, since the sacs containing the glands, and presumably secreted by the glands themselves at an early stage, consist of a refractive membrane, while the lumen of the fronto-lateral horns is often partially interrupted by septa of apparently similar nature and formed by the adhesion of the membrane of the globules to the walls of the horns (see p. 174).   The chitin, however, must be of a delicate nature, for while resembling the cuticular covering of the whole body in being soluble in warm acid (HNO_3), it also dissolves in hot caustic potash, which the ordinary cuticle resists.

It is worthy of note that the period of secretion of all three glands is almost coincident with that of the free life of the larva, the secretion passing to the exterior after the first moult, i.e., shortly after hatching, and ceasing when the Cypris-stage becomes fixed, but occasionally while free.   It is possible that no new secretion is formed during the Cypris stage, and that the glandular material present is simply a relic of that produced during the earlier stages.   The period of secretion would then be that during which the larva feeds, for, as is well known, the Cypris form does not feed.

The presence of sharp points at the ends of the fronto-lateral horns in the earlier stages, and of a strong spine projecting from the horn during the later stages seems, as CLAUS and HOEK have already pointed out, to indicate that they may be piercing organs provided with poison glands.   It is to be observed that the area covered by the horns is that included by the sweep of the appendages, and that any organism paralyzed by the secretion would tend to be swept towards the region of the mouth.

The lateral gland and the other glands of the carapace of later stages appear from their position more adapted for protection than for securing prey.

The position of the gland on the labrum and the presence of a perforated tooth for the passage of its secretion, taken in conjunction with the supposed free motion of the distal lobe of the labrum, indicate that it may be used to pierce and paralyze organisms held by the stiff spines on the endopodites of the antennæ and mandibles in the way supposed in the preceding section.

(D.) EXCRETION.

With the object of determining whether any portions of the body were specialized for secretion, the Nauplii of *Balanus perforatus* were fed with powdered carmine, indigo-carmine, Bismark brown, and methyl-blue in sea-water. All these substances were readily taken up, but, though the carmine and indigo-carmine formed large accumulations in the stomach and intestine, none appeared elsewhere; nor is it probable that the alimentary canal has an excretory function in the way found by CLAUS for Copepods (58), as colouring matter did not occur in any of the cells of the digestive tube. The Bismark brown on the other hand appeared to be digested and excreted again all over the body, so that ectoderm, muscles, and nerve cells, etc., appear to have a certain excretory power. With methyl-blue more definite results were obtained : the colouring matter is excreted in the ectoderm as small blue granules which give a blue tint to the whole dorsal surface, the carapace especially, of the Nauplius ; the excreted granules also extend into the base of the first and second pair of appendages, and are especially numerous in the epithelium of the labrum ; they are also abundant in the mesodermal cells covering the stomach and scattered about on the muscles. They generally occur in circular patches (figs. 148, 148a), corresponding with the protoplasmic accumulations. In order to make certain that the particles were not taken up by the ectoderm, Nauplii which had fed on the colouring matter, and had a considerable accumulation in the stomach, but none in the ectoderm, were isolated and placed in fresh sea-water; after a lapse of some time the methyl-blue was seen to be excreted precisely as before.

It would thus appear that the lining of the body cavity has a special excretory function, more particularly in certain regions.

The Nauplii of *Lepas anatifera* and *Chthamalus stellatus* excreted methyl-blue in the same manner.

No excreted particles were found in the fronto-lateral or lateral glands, or in the axial glands of the labrum.

(E.) PERCEPTION.

The frontal filaments from their intimate connection with the brain are no doubt sensory organs. Their position and ventral direction would render them well adapted to test the chemical nature of the substances with which they came into contact before reaching the mouth, and the term " olfactory filaments " applied to them first by FRITZ MÜLLER is, no doubt, suitable.

The perception of light by the Nauplii of most Cirripedes · examined is very marked, but whether this takes place over the general surface as in some animals (78), or is limited to the Nauplius eye or some other spot there is no evidence to show.

(F.) EFFECT OF STIMULI.

The reaction to light is usually very marked and has been already treated of at length by Dr. LOEB and myself. It will be sufficient to state here that during motion the Nauplii of *Balanus perforatus* commonly tend to place their longer axis parallel to the rays of light ; that light of sufficient intensity and duration ordinarily causes them to turn the oral pole away from the light, while weaker light has after a time the contrary effect. For fuller details as to the action of the rays of different refrangibility, and of sudden increments or diminutions in the intensity of light, I must refer the reader to the original communication. We have there shown that the effect of this must be to produce the daily, and possibly also the annual, migrations to and from the surface, such as have been found to take place in the Nauplii of *Lepas fascicularis* (28) and many other pelagic forms.

The wide prevalence of this effect of light on animals (and even on plants), as shown by the movements of such pelagic forms (whether fresh water or marine), hemi-pelagic forms, and larvæ which are not pelagic forms, which are so different that we must suppose the effect to have arisen independently in the different groups, appears to indicate the great physiological importance of the periodic movement. Whether this is to regulate the amount of intensity of light received, or has some other object is not clear.

The immediate effect of an increase of heat above the normal winter temperature appeared to be at first a great increase in the activity of movement, but after a certain temperature was reached the Nauplii became less active, and if the temperature was not then lowered died ; if the heat was again diminished an increase of activity took place. The maximum, optimum, and minimum temperatures for the activity of the Nauplii were not determined.

(G.) EFFECT OF GRAVITY.

Gravity has apparently a definite effect on the Nauplii of *Balanus perforatus* (and possibly on those of other species, but these were not examined in this respect). If they are very closely examined with a lens or low power of the microscope it is seen that the Nauplii generally move along with their ventral surface upwards ; this is specially clear in the case of the larger and older Nauplii, but in consequence of their small size and rapid motion is difficult to make certain of in the younger. The circumstance has apparently no relation with the surface of the water or with the bottom of the vessel, as it occurs in whatever part of the vessel the larvæ are.

PART V.—GENERAL CONSIDERATIONS.

(A.) SEGMENTATION OF THE OVUM.

The segmentation of the ovum has been described by the best observers as total, followed by epiboly. The migration of the protoplasm to one pole does not, however, as has been generally supposed, represent simply a division into epiblast and hypoblast ; since the supposed representative of the latter (the yolk) is at first devoid of a special nucleus, and furnishes, moreover, at a slightly later period material

2 c 2

for the production of new epiblast cells. The process is probably simply the formation of a telolecithal egg.

Fig. 3.

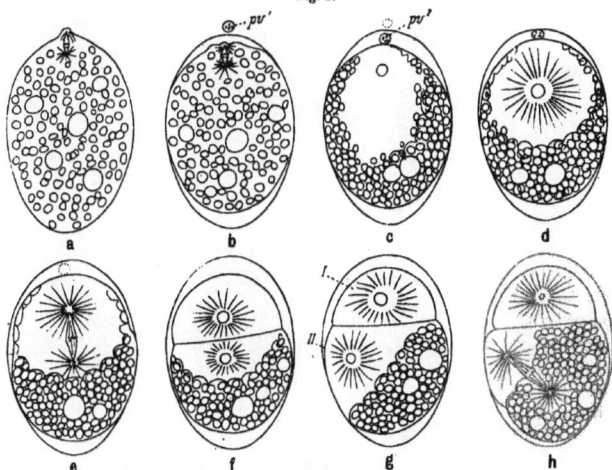

a...s. Diagrams illustrating the mode of segmentation of the Cirripede ovum, the formation of the blastoderm, and of the three germinal layers. The protoplasm is left white; the yolk is granular, with oil-drops. I, II, and III, the three first blastomeres, formed directly from the yolk; pv.¹, first polar body; pv.², second ditto. (The stellate figures at the ends of the directive spindles in figs. a and b were inserted through inadvertence.)

In a, the formation of the first polar body is shown; in b, that of the second, from the contractile ovum now withdrawn from the newly-formed vitelline membrane; in c, the protoplasm is collecting at one end, and the segmentation-nucleus is becoming conspicuous; in d, the segmentation-nucleus is preparing to divide; in e, the protoplasm is collected at the anterior end, and the nucleus is dividing; in f, the first blastomere has been cut off from the yolk; in g (lateral view), the second blastomere is forming; in h, the nucleus is dividing to form a third blastomere; in j′ (side view) and j″ (front view), the second blastomere is cut off from the yolk, and the first has divided transversely; in k (front view), the second is likewise dividing transversely, and the nucleus of the third giving rise to a new one; in l, II has completely divided, and III is cut off from the yolk; in m, the blastoderm now covers a good part of the yolk, a new blastomere is forming, and the nucleus is dividing, one daughter-nucleus being in the yolk; in n, the blastopore is being closed by the emergence of a merocyte; in o, this is dividing to form the last blastodermic cell, and complete the epiblast; in p, the blastoderm is

The segmentation is diagrammatically represented by the woodcuts fig. 3, a, b, c, &c. Figs. c and d show the concentration of the protoplasm at the anterior end. As the first blastomere becomes cut off from the yolk the nucleus divides (e, f) and one daughter-nucleus passes into the yolk half, and soon emerges (*g*) accompanied by protoplasm to form a second blastomere and generally situated close to the first.   As

Fig. 3—(continued).

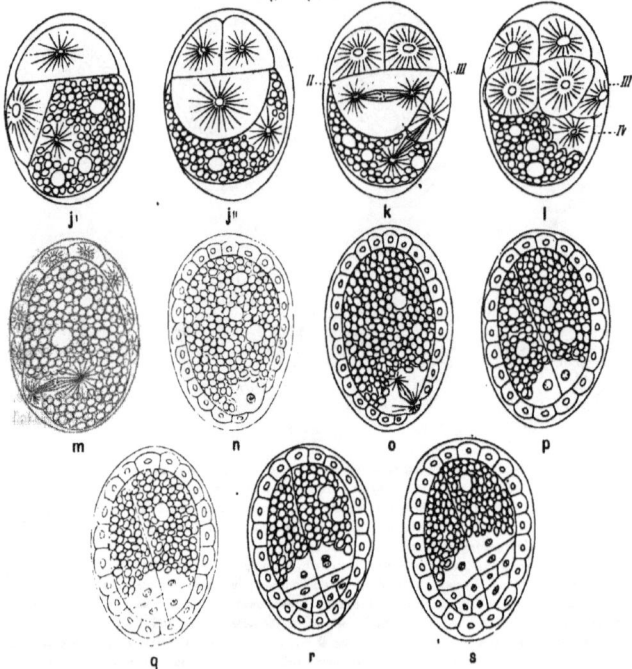

completed, and the nucleus left in the yolk has divided in two, and with it the whole yolk; in *q*, the two yolk-cells thus formed have cut off two mesoblastic cells at their hinder ends; in *r*, these latter have divided, and two new mesoblastic cells have been cut off from the yolk-cells; in *s*, the germinal layers are complete, the mesoblast occupies the hinder end, and the two yolk-cells, with their posterior nucleated protoplasmic masses the anterior end; the epiblast covers the whole.   (*m*...*s* are side views.)

this becomes cut off from the yolk it gives off (*h*) into the yolk a nucleus, which, behaving (*j, k, l*) similarly to the daughter-nucleus of the germinal vesicle, forms new protoplasm and emerges as a third blastomere.   At each successive stage the yolk is in communication with one merocyte or newly-forming blastomere, and this, before becoming shut off as a blastomere, gives off a single nucleus into the yolk.

The yolk may be, it will be seen, regarded as a single cell with abundant nutritive material (and at times very little protoplasm), the nucleus of which by its division and creation of new protoplasm from the yolk granules forms new cells at its surface, and moves about from place to place in order to invest the yolk with a covering of blastomeres : the position of the nucleus (or its spindle) always more or less peripheral, and the idea of a central nucleus giving off nuclei radially, would not quite meet the case.

The mode in which the blastoderm grows round the yolk agrees essentially with the process of epiboly as defined by LANG (79). The definite endoderm is not formed till late, but the yolk may be regarded as a generating cell giving off in succession epiblastic cells which themselves divide up and tend to complete the epiblastic investment.

The circumstance that the yolk thus represents a single cell will exclude, I think, the view suggested by KORSCHELT and HEIDER (81) that the Cirripede ovum segments superficially during the later phases of cleavage.

This process may rather be compared to that taking place in the case of *Bonellia*, the four macromeres of this type being represented by a single one in the case of Cirripedes.

It is worthy of note that the yolk never appears to be cut off from all communication with the protoplasm in the way stated by some observers in the case of certain other Crustacea ; one of the peripheral micromeres, or one of the merocytes derived from the latter, is always in communication with it, and it is, therefore, not to be regarded as extra-cellular. Nor is it to be regarded as belonging, from the first, solely to the hypoblast, for until the closure of the blastopore epiblastic cells are formed at its expense, and afterwards for a time also the mesoblast.

The effect of the nucleus in the transformation of yolk material into protoplasm is very marked in the Cirripede ovum. The yolk itself, throughout the main portion of its mass, contains very little protoplasm. All stages can be traced between a condition in which a nucleus, only recognizable by a slight stellate arrangement of the smaller granules with which it is surrounded, occurs near the micromere or emerging merocyte from which it has apparently been given off, to a well-formed micromere with abundant protoplasm. The radial arrangement becomes more definite by the continued transformation of the large yolk granules and oil globules into granular protoplasm, the granules of which become more and more distinctly arranged in radial rows : the area of transformation gradually extends, and eventually a considerable section of yolk is bodily transformed into protoplasm. When once this is seen it is very evident how all the primitive micromeres have arisen in this manner. The nucleus, accompanied as it is in most cases by little or no protoplasm as it passes into the yolk, evidently possesses this transforming power.

In describing the details of division of the cells of the blastoderm and yolk-endoderm much variation has been shown to occur, so much indeed that the process may be termed irregular. Such differences show well the morphological insignificance of the details of cell division in the present case, for the Nauplii vary proportionately much less; every one of the numerous, simple, or compound bristles or spines of the Nauplius has its definite character and position, which are maintained with surprising constancy throughout, although they must have been produced by epiblast cells having very different modes of origin and arrangement.

(B.) DIFFERENTIATION OF THE GERMINAL LAYERS.

With respect to the origin of the mesoblast and hypoblast of the Nauplius, the Cirripedes occupy an isolated position among Crustacea.

They differ from *Cetochilus* (35) among Copepods, *Moina* (65) among Phyllopods, and from *Astacus* (73), and most other Decapods, as well as from *Ligia* among Isopods (81), in the fact that the hypoblast and mesoblast do not arise as multicellular areas forming a specially differentiated and separate portion of the blastoderm, but from a common source, the single yolk cell, the nucleus of which is derived from the last, or possibly, in rare cases, from one of the last formed cells completing the blastoderm and filling the blastopore.

The mode in which the mesoblast and hypoblast are differentiated, presents most resemblance among Decapods to that found in *Palæmon* and *Eriphia*, where the mesoblast arises from the invaginated portion of the blastoderm, which also forms the mid-gut (and hind-gut). In *Oniscus* and *Cuma*, too, a simple blastodermic proliferation gives rise to both mesoblast and hypoblast, and the same appears to be true of *Daphnia* (81). The case of *Cyclops* may also be mentioned in this connection. Here the endoderm is stated to be derived by invagination (64), or from a single cell (67), which also possibly gives rise to the mesoderm; the mesenchyme cells, however, which give rise to the muscles of the Nauplius, are stated by URBANOWICZ to be derived from the ectoderm. The process in these cases may be regarded as representing an invagination of the meso-hypoblast, and that seen in Cirripedes may be looked upon as the same process reduced to its simplest expression, viz., a single cell (the uni-nucleated yolk) representing the common origin of both mesoderm and endoderm.

This cell divides immediately into a more dorsally and a more ventrally situated cell; from the hinder end of each of these cells uni-nucleated segments are cut off to form the mesoblast. When this process is completed, the yolk still consists of two cells; these represent the whole of the endoderm; behind this is a plug of mesoblast cells. The hypoblast then, at the time of its differentiation from the mesoblast, consists of two cells, while the mesoblast is multicellular.

The two protoplasmic bodies of the yolk-endoderm cells may be compared with the amœboid cells at first bounding the archenteron in *Palæmon*, and probably with the cells with processes radiating into the yolk, and situated at the end of the invaginated portion of the blastoderm in *Eupagurus*.

The mode of origin of the mesoblast clearly shows that no paired pole-cells (teloblasts) such as have been supposed by GROBBEN to give rise to the mesoblast in *Balanus* and *Peltogaster* (35), are present.

No reproductive cells, such as are described in *Moina* and *Cetochilus*, can be recognized at an early period in Cirripedes.

In the circumstance that the blastopore closes completely, Cirripedes agree with *Moina*, *Cetochilus*, and *Cyclops*; it differs, however, possibly in position from that of *Moina*, where it is supposed, by GROBBEN, to close on the site of the future mouth. It resembles that of the majority of Crustacea in showing no extension in the direction of the mouth.

(C.) FORMATION OF THE ALIMENTARY CANAL.

The yolk-endoderm gives rise to the stomach (with its glands) alone of the Nauplius and adult Cirripede.   As the mesoblast develops into muscles and connective tissue, the central part of the yolk is absorbed, the nucleus retiring to the periphery ; in this way the hollow mesenteron is formed, the walls of which are finally composed of a number of clear nucleated endoderm cells devoid of yolk, each formed by the centrifugal contraction of a yolk-pyramid.   The yolk-pyramids are essentially similar in structure, relations, and fate to the secondary yolk-pyramids of *Astacus* (REICHENBACH'S "Secundäre Dotterpyramiden"), and are evidently the equivalents of the latter.

(D.) ORIGIN OF THE NAUPLIUS APPENDAGES.

A striking feature of the mode of origin of the Nauplius appendages is that they appear first on the dorsal side of the embryo.   It is only, however, the free ends which are thus seen, the main part of the appendage being applied to the sides of the body, and the origin, as usual among appendiculate animals, ventral.

In Cirripede groups other than Thoracica a similar mode of origin occurs : VAN BENEDEN (58) has already described the same thing in *Sacculina*, and I have observed it myself in *Peltogaster*.   A similar disposition of the Nauplius appendages in *Laura* has caused LACAZE-DUTHIERS (38) to state that the appendages are attached on the dorsal side, a peculiarity he specially emphasizes.   The attachment is, I believe, ventral as usual, but the free ends are dorsally directed, the position of the embryo having been inverted by this observer.

It seems very probable that the same fact obtains in the Nauplii of all the other groups.

In the Phyllopods the figures of GROBBEN make this clear, and he states, for *Moina*, "Sowohl die beiden Antennen als die Mandibel wachsen von innen nach aussen."

In *Cetochilus* the same author failed to follow the origin of the appendages, but says, "Nachdem sich die auf die Dorsalseite zurückgelegten Extremitäten deutlicher entwickelt haben. . . ."

In *Anchorella*, *Lernæopoda*, and *Brachiella*, VAN BENEDEN (60) states that the Nauplius appendages develop from within outwards.

The same is apparently true of *Tracheliastes polycopus* (63) and *Argulus foliaceus* (62), as well as *Nebalia* (68).

In *Mysis ferruginea*, VAN BENEDEN (70) figures and describes a similar disposition of the Nauplius appendages.

In Decapoda, Amphipoda, and Isopoda, where the Nauplius stage is represented but concealed, the Nauplius appendages likewise grow dorsally, as follows from the descriptions or figures of REICHENBACH (73), FAXON (72), KINGSLEY (74), and others, for the Decapoda (*Astacus, Palæmonetes, Crangon*), and from the figures of RATHKE (75) for Isopoda and Amphipoda (*Bopyrus, Amphithoe*), and of VAN BENEDEN for *Asellus* (76).

This mode of development holds good, therefore, in all probability, for the Nauplius or Nauplius-stage of Phyllopoda, Cirripedia, Copepoda, Leptostraca, Schizopoda, Decapoda, and Thoracostraca (Amphipoda, Isopoda).

As early as 1869, VAN BENEDEN indeed drew attention to the generality in the difference in the mode of development of the Nauplius appendages and the later ones, and, speaking of *Asellus*, says, "Ses organes se développent rapidement, et s'allongent en se portant en arrière et en dehors. Il parait en être de même chez tous les Crustacés. Partout les antennes et les mandibules semblent se développer de dedans en dehors, et, par là, les trois premières paires d'appendices se distinguent de tous les autres, qui se développent, au contraire, de dehors en dedans, en se rapprochant de la ligne médiane."

This law, however, has been generally overlooked, not only in Cirripedes, but also in Copepods; and it appears almost certain that in the latter group, as well as in the former, the surface of the embryo, on which the median longitudinal and transverse furrows appear, and which has been described as ventral, is in reality dorsal.

The dorsal position of the mesoderm plate is evidently in relation with the dorsal position of the appendages, for a considerable part of the mesoderm goes to form the latter. URBANOWICZ (67) describes a similar accumulation of his "mesenchyme" cells on the dorsal side in *Cyclops*, and states that they form the muscles of the appendages.

(E.) BODY CAVITY.*

Inasmuch as the body cavity arises partly owing to the more rapid growth of the ectoderm and the endoderm, and consequent separation of these layers, it might be

---

* This term is used in a purely descriptive sense, to denote generally the space included within the body walls.

termed a blastocœle; but since mesoderm cells exist between the two layers, and
upon separation of these remained, some associated with the ectoderm, some with the
endoderm, and some spanning the space intervening between the two, it may be said
with equal truth to be a schizocœle, and regarded as arising by a failure of the mesoderm to keep pace with the growth of the layer bounding it, and its consequent
splitting or excavation by the formation of vacuoles. If no mesoderm were present at
the stage when the cavity arose, it could be termed a blastocœle; if a complete mesodermic layer were present and split into somatic and splanchnic layers, it would be
termed a schizocœle; but since this layer is not complete, and a true split does not
occur, the body cavity may be said to be in part a schizocœle and in part a
blastocœle.

The mesoderm of the Nauplius shows no trace of an arrangement into somites,
and the body cavity is continuous from one end of the body to the other; the mesoderm thus resembles in its deportment that of the cephalic extremity in *Astacus*.

(F.) NERVOUS SYSTEM.

The nervous system arises, probably, mainly or altogether as an epiblastic thickening, and at Stage 2 retains everywhere its primitive connection with the ectoderm;
it has ganglion cells along its whole extent.

In other respects the nervous system shows much specialization. It is sharply
marked off from the rest of the ectoderm, and is distinctly divided into ganglia and
connectives. The commissures are practically absent, and so far from showing
ganglia corresponding to the metameres indicated by the appendages, the nervous
system is considerably complicated, even at this early period. It shows, however,
an anterior portion, which is in close connection with the frontal filaments and
Nauplius eye, and may be regarded as corresponding to the archi-cerebrum of
LANKESTER.

In Balanids two accessory and a central lobe of unknown significance are present,
as well as short thick nerves passing to the labrum.

The two accumulations of ganglion cells situated just at the commencement of
the circum-œsophageal connectives, and coming into close relation with the antennules, probably correspond to the ganglia of the first post-oral somite, and inasmuch
as these cells belong to the brain, the latter may be said to be a syn-cerebrum from
the first, though its component elements, the archi-cerebrum and ganglia of the
antennules, may still be recognized.

The ganglion cells on the circum-œsophageal connectives which come into close
relation with the origin of the antennæ probably represent the ganglia of the second
post-oral somite, while the large sub-œsophageal ganglion represents the fused
ganglia of the mandibles.

(G.) MORPHOLOGY OF THE APPENDAGES.

Comparing the three pairs of appendages, the antennæ and mandibles are seen to be very similar in structure. Each consists of a two-jointed protopodite bearing inwardly directed plumose spines ; the endopodite has a series of stiff bristles or spines forming a series continuous with those of the protopodite ; the bristles are given off generally in pairs, while the distal end of each joint of the much more jointed exopodite gives off a single bristle.

The single ramus of the antennules resembles the protopodite and endopodite together more closely than it would the protopodite and exopodite, the joints being few and the bristles given off singly, or in groups of two or four ; one bristle is given off from the middle of the third joint like the two or three on the first joint of the endopodite of the antennæ and mandibles.

It is therefore probable that the appendages are serially homologous, and that it is the exopodite of the antennules which is absent.

That the antennules are not of a different nature to the two remaining pairs of appendages is also indicated by the similar and peculiar origin of all three.

I believe, therefore, that all the appendages are of the same kind, and may represent, as LANKESTER contends, for Crustacea generally (66) primitively post-oral appendages.

It is important to remark that the first two pairs of appendages are, however, never ontogenetically post-oral, the antennules being from the first in front of the mouth, and the antennæ at its sides ; this, however, together with the pre-oral innervation, may be one of the many signs of early specialization visible in the Nauplius, such as have led to the view that the Nauplius is a trochosphere with precocious Crustacean characters (70).

(H.) COMPARISON OF THE NAUPLII OF THE DIFFERENT SPECIES.

(α.) *Points of Agreement.*

In the general description it has already been seen how closely the Nauplii of all Thoracic Cirripedia resemble each other. Of these points of agreement we may distinguish—

(i) Characters shared by all the Nauplii.

(ii) Special characters of the Cirripede Nauplius.

Though until the Nauplii of the various sub-divisions of Crustacea have been properly examined it will be impossible to assign all the characters to one or other of these divisions, the striking features of the Nauplius are well known, and need not be repeated here.

Among the peculiar features of the Cirripede Nauplius the following are the most important :—

1. The shape of the carapace with its fronto-lateral horns and caudal spine.
2. The presence and structure of the fronto-lateral glands.
3. The size, shape, and structure of the labrum.
4. The structure of the axial gland of the labrum.
5. The character of the setose region.
6. The structure of the appendages.
7. The size and form of the tail (thorax-abdomen).

The agreement of the Nauplii of genera so distinct as *Lepas* and *Balanus*, as well as of Nauplii more or less intermediate between these types, such as *Chthamalus*, in these particulars is highly remarkable and significant.   The perfect similarity which obtains in all the species I have examined in the number, disposition, and minute character of the very numerous bristles and other processes on the appendages of Nauplii of the second stage—a similarity which cannot be supposed to be merely analogical—demonstrates that the character of the appendages is a primitive one, actually possessed by the common ancestor of the Thoracica at some stage in its life history.

The majority of the characters common to all the Nauplii being shown in this way to be ancestral, it becomes necessary to determine whether the Nauplius of the Thoracica represents in any sense their ancestor, the former adult characters being supposed to have been precociously acquired by the larva, or whether the features in question belonged to a Nauplius possessed by the ancestor.

Now the structure of the adult *Lepas* and *Balanus* is so obviously similar that we can only suppose the two to have diverged from a similar ancestor.   The ancestral adult did not therefore possess the peculiar features of the Nauplius.

The correspondence between the two forms at other stages is so close that we can only suppose the essential features of each stage to be ancestral; thus the segmentation of the ovum, the peculiar mode of formation of the embryo—so similar (as I have shown) in all the species—must have been shared by the ancestor of the group.   Of the Cypris stage DARWIN says (10) : " In the pupæ of all these genera (*Lepas, Conchoderma, Dichelaspis, Ibla, Alcippe,* and *Balanus*) there is a most close general agreement in structure, excepting in minute detail.   I was surprised to find exactly the same slight differences in the spines on the first pair of natatory legs, as compared with the succeeding pairs, in *Balanus hameri* as in *Lepas*."

We cannot escape from the conclusion that the very similar development in *Lepas* and *Balanus* (which I hope to illustrate in a further communication)* corresponds stage for stage with that of the ancestor of the Thoracica, or in other words, that the

---

* I may state that before reaching the Cypris stage *Balanus perforatus* undergoes five moults; *Lepas fascicularis* probably undergoes the same number, though owing, apparently, to inadvertence the number is said by WILLEMOES-SUHM (28) in one place to be seven.   The six Nauplius stages correspond, probably, exactly.   The agreement of the first two stages in *Balanus, Chthamalus, Conchoderma,* and *Lepas* in minute detail has been shown above.

latter underwent a metamorphosis perfectly similar to that of the present members of the group.

It will not be difficult to test this conclusion, for a careful study of the affinities of the Thoracic Cirripedes indicates that the Balanids, on the one hand, have probably diverged from *Pollicipes*, and the majority of the Lepads, on the other, from *Scalpellum*; these two genera are connected, as DARWIN and HOEK have shown, by intermediate forms. The ancestor of the whole group has therefore been preserved probably in a very slightly altered form, and it will be possible to study its development in detail. As far as an approach has been made to this, the results confirm the above conclusions; *Scalpellum* apparently, from the descriptions of HESSE (12, 25), passes through a metamorphosis similar to that of *Balanus* and *Lepas*, and the segmentation of the ovum, as figured by LANG in *Scalpellum*, and NUSSBAUM in *Pollicipes*, does not differ from that of the other Thoracica, while the larva in the latter genus is a typical Thoracic Nauplius.

The permanence of such minute characters as the arrangement of the bristles on the appendages for the vast time represented by the Tertiary, Cretaceous, and probably, at least, part of the Jurassic periods, is highly remarkable, and well shows the slow rate of evolution which may take place in so highly specialized a group.

(β.) *Larval Differences.*

In spite of the great agreement of the different Nauplii in many points, there are nevertheless well-marked differences by which the Nauplii of all the genera examined, and often of the species, can be readily determined.

These points of difference relate to—

1. Size.
2. Shape of the carapace.
3. Length of the fronto-lateral horns.
4. Length and character of the caudal spine.
5. Shape and size of the labrum.
6. Form of the axial gland of the labrum.
7. Character of the setose region.
8. Slight modifications in the structure of the appendages, and in the relative proportions of their parts.
9. Length and character of the tail.
10. Extent of "ventral plate."
11. Structure of the nervous system.
12. Presence or absence of lateral glands.
13. Presence or absence of certain connective tissue elements.
14. Physiological properties—Heliotropism—Rapidity of motion.

In the following Tables a classification of the Nauplii according to their resemblances, and perfectly independent of conclusions drawn from the adult, is given.

(γ.) *Classification of Cirripede Nauplii of Stage* 2.

Lepas, Conchoderma, and *Dichelaspis* (Lepadidæ).

 i. Nauplii long (0·6 to 0·8 millim.), slender and transparent; movements slow (at least in the first two genera).
 ii. Caudal spine long and covered with secondary spinelets.
 iii. Labrum long; proximal lobe ovate, distal relatively large and pentagonal; axial gland well developed in both divisions of labrum.
 iv. Tail long, with secondary spinelets; ectodermal thickening represented by two small groups of cells.
 v. Brain two-lobed (accessory lobes absent).
 vi. Lateral glands absent.
 vii. All the bands of the setose region present (*Dichelaspis ?*).

(A.) *Lepas* and *Conchoderma*.		(B.) *Dichelaspis*.
(α.) Carapace large, shield-shaped.		(α.) Carapace narrow, triangular.
(β.) Fronto-lateral horns long and slender.		(β.) Fronto-lateral horns of moderate length.
(γ.) Distal lobe of labrum broad, with one large median and two lateral teeth; axial gland narrow distally.		(γ.) Distal lobe of labrum narrow, with one large median tooth; axial gland dilated distally.
(δ.) Maxillary band of setose region well developed.		(δ.) Setose region ?
(ε.) Ova blue, relatively broad.		(ε.) Ova vermilion-red, narrow (*D. Darwinii*).
Lepas.	*Conchoderma.*	
Tail simple.	Tail bifid.	
L. *fascicularis,* length 0·6 mm. L. *pectinata* „ 0·7 „ L. *anatifera* „ 0·8 „		

Balanus and *Chthamalus* (Balanidæ).

 i. Nauplii of moderate length (0·23 to 0·46 millim.); movements rapid.
 ii. Caudal spine of moderate length, with very short secondary spines; fronto-lateral horns short or of very moderate length.
 iii. Labrum shorter and broader than in Lepadidæ; proximal lobe broad; distal lobe relatively small and rounded; axial gland developed almost exclusively in the distal portion (*Chthamalus ?*).
 iv. Tail short.

v. Brain complex (accessory lobes present).
vi. Lateral glands present.
vii. Maxillary band of setose region as yet absent.

(A.) *Balanus* (Balaninæ).	(B.) *Chthamalus* (Chthamalinæ).
(α.) Carapace shield-shaped, with two small lateral teeth.	(α.) Carapace almost circular.
(β.) Tail and caudal spine with minute teeth; ectodermal thickening of tail a considerable plate on each side.	(β.) Tail and caudal spine with strongish spinelets, especially at sides of base of latter; ectodermal thickening of tail a few cells on each side.
(γ.) Proximal lobe of labrum with two small lateral lobes; distal lobes small, with two lateral teeth.	(γ.) Proximal lobe of labrum simple; distal with two lateral, several central, and smaller teeth.
(δ.) Pre-maxillary bands slight; anterior band complex; extra-maxillary arc slight.	(δ.) Pre-maxillary bands strong and united together; anterior band and post-oral group ?; extra-maxillary arc distinct.
(ε.) Ova brown; yolk granules large; oil globules small.	(ε.) Ova orange coloured, narrow; yolk granules small; oil globules large.

B. *perforatus*.	B. *improvisus*.	B. *balanoides*.
Length 0·46 mm.	Length 0·27 mm.	Length 0·45 mm.
Horns moderately short.	Horns moderately short.	Horns short.
Tail rather short.	Tail rather short.	Tail short.
Caudal spine rather short.	Caudal spine rather short.	Caudal spine longish.

These tables illustrate the differences which may obtain among the larvæ of allied animals.

We see that, among Cirripedes, the larvæ and embryos of all the genera differ not inconsiderably. The embryos of *Lepas*, *Conchoderma*, or *Balanus* can be distinguished at any stage by one or more features, and the Nauplii much more readily. The species, or even in some cases the genera are not so readily recognized, for though, in *Lepas anatifera* and *L. pectinata*, the ova and embryos can with care be distinguished by the size and shape, yet the same stages of *Conchoderma virgata* and *Lepas anati-fera* can hardly be distinguished at any stage, and even the Nauplii can only be separated by the slight fork in the tail of the former.

The differences between the species go back as far as the new laid ovum, as may be seen from the sections on the size, shape, colour, and constitution of the ova. Thus the narrow orange-coloured ova of *Chthamalus* with their large oil drops may be distinguished from the broader yellowish ones of *Balanus* with their more numerous but smaller oil globules, and from the smaller vermilion eggs of *Dichelaspis*, and from the broader blue ova of *Lepas* and *Conchoderma*. The ova of *Lepas pectinata* are narrower than those of *L. anatifera* and *Conchoderma virgata*. In the genus *Balanus* the ova of the three species *B. improvisus*, *B. perforatus*, and *B. balanoides* differ much in size, the first-mentioned having the smallest and the last the largest ova, measurements of which have been given. Sections of *Lepas* and *Balanus*, and probably of *Chthamalus*, at any stage of embryonic development can be distinguished by the characters of the yolk. It must be stated, however, that the specific and even generic differences (*cf. Lepas anatifera* and *Conchoderma virgata*) may not be recognized in single specimens, and that it may be necessary to compare a number of cases in order to eliminate the effect of variation.

The Nauplii at the second stage can be readily classified and might, in most cases, be used for the formation of families, genera, and species. The differences certainly are not very great, but seem to be as important relatively to the Nauplii as the generic differences are to the adult.

It is instructive to observe that the classification of the genera as deduced from the developmental stages agrees perfectly with that arrived at from the relations of the adult: the differences between the Nauplii of the different genera are approximately proportional to those between the corresponding adults; thus the two genera most widely remote from one another, *Lepas* and *Balanus*, also show the greatest differences in their Nauplii. The Chthamalinæ are those members of the Balanidæ which approach the Lepadidæ most closely, and their Nauplii show similar affinities. *Lepas* and *Conchoderma* are perhaps, in some respects, the highest members of the Lepadidæ, and their Nauplii show, in some respects, more special features than those of *Dichelaspis*, which branched off from the main stem at an earlier period. The two former genera so closely allied in the internal organization have likewise Nauplii so much alike as to be almost indistinguishable.

(δ.) *Larval Evolution.*

The phylogeny of the Cirripedes, as deduced from the adult anatomy, may be represented as follows :—

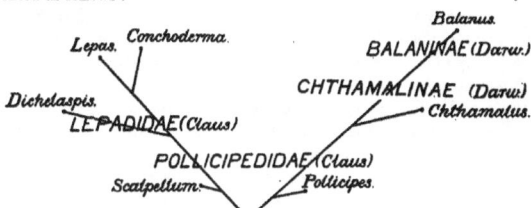

The Nauplii fit in well with this scheme; they have undergone an evolution parallel to and simultaneous with that of the adults, and the results obtained in each case are nearly proportional.

It is a point of some interest to inquire whether the new characters assumed by the larvæ have been acquired independently, or are due to a precocious appearance of features properly belonging to the adult, in the way indicated by FRITZ MÜLLER (16) and WEISMANN (79), or whether both processes have taken place.

Glancing down the list of larval differences given on pp. 205–7, it is evident that though some of the characteristic differences may possibly be due to precocious acquirement by the young Nauplius of the forms in question of characters originally common to older larvæ at an earlier phylogenetic stage, such as the early acquirement of glands (the lateral glands) in Balanids possibly properly belonging to the Archizoœa stage in Lepads, yet few of the characters can be conceived as transferred back from even a young sessile form; most of the differences affect structures peculiar to the Nauplius, and lost by the adult.

It is thus evident that the variation in the Nauplii (the same may be said of the ova and embryos), though always accompanying adult variation, has taken place in a perfectly distinct direction. Differences in the length of the horns, in the length and armature of the caudal spine, in the character of the setose region, or in the presence or absence of the lateral glands must have been produced in this way.

Few probably of the other characters can be referred to larval precocity. Though the species which have the smaller sized Nauplii have as a rule the smaller sized ova* (see table, p. 130), yet in neither case is the size proportional to that of the adult; thus, though the Nauplius of *Lepas anatifera* is larger than that of the smaller *L. pectinata*, the latter has a Nauplius larger than that of *L. fascicularis*, which is about twice the size. Similarly, the small *Balanus balanoides* has a Nauplius nearly twice the size of that of the perhaps rather larger *B. improvisus*. The labrum of the Nauplius, with its axial gland, is practically lost at the Cypris stage, and that of the sessile Cirripede differs considerably from the earlier structure. The greater complication of the brain in the Balanids may be a precocious character, but until the brain of the Nauplius and of the adult have been compared this, too, is uncertain. The early appearance of the thickening on the tail, which gives rise to important parts of the body of the thorax in *Balanus*, is the only character which appears definitely to indicate the precocious appearance of features originally belonging to a later ontogenetic stage.

The origin and meaning of the Nauplius hardly come within the scope of the present subject, since, as MARSHALL (80) has pointed out from other considerations, the origin of this larval form is outside the group.

I must, in conclusion, express my best thanks to Messrs. SEDGWICK, HARMER, and WELDON, at whose suggestion I undertook the study of the Cirripedes, and who have

---

* The shape, too, is to a certain extent adapted to that of the Nauplii, *cf. Lepas, Dichelaspis,* and *Balanus.*

2 E

frequently assisted me with their advice; to Professor WELDON I am specially indebted for many valuable criticisms. My warmest thanks are also due to Dr. EISIG, who showed me much kindness, and assisted me in every way possible during my stay at the Zoological Station at Naples. To Dr. PAUL MAYER I am indebted for the specimens of the Nauplii of *Dichelaspis*, and for information kindly given me on certain of his methods. To Professor KOWALEVSKY I am indebted for suggestions as to the best method of investigating the process of excretion in the Nauplii. To my friend Dr. HOEK for the kind way in which he has assisted me by the gift or loan of literature treating of Cirripedes; and lastly, I have to thank Signor LO BIANCO for the ready way in which he provided me with the large amount of material necessary for embryological study, and Professor MIALL for affording me facilities for completing my work in the laboratory of the Yorkshire College.

. ʹɾͼ **Yorkshire** College,
ʰ͵ ʹɗͼ,   PART VI.—BIBLIOGRAPHY.

(1.) MARTINUS SLABBER.   Naturkundige Verlustigungen.   1778.

(2.) J. V. THOMPSON.   Zoological Researches and Illustrations, vol. 1, Part 1. Memoir 4.   On the Cirripedes or Barnacles.   Cork.   1830.

(3.) J. E. GRAY.   On the Reproduction of Cirripedia.   Zool. Soc. Proc., vol. 1, 1833.

(4.) H. BURMEISTER.   Beiträge zur Naturgeschichte der Rankenfüsser.   Berlin. 1834.

(5.) J. V. THOMPSON.   Phil. Trans., Part II., 1835.

(6.) KOREN and DANIELSSEN.   Zoologiske Bidrag.   Bidrag til Cirriperdernes Udvikling.   Nyt Magazin for Naturvidenskaberne.

(7.) H. GOODSIR.   Edinb. New Phil. Journal, vol. 35, 1843.

(8.) C. SPENCE BATE.   On the Development of the Cirripedia.   Annals and Mag. of Nat. Hist., Second Series, vol. 8, 1851.

(9.) C. DARWIN.   A Monograph of the Sub-class Cirripedia, vol. 1, Lepadidæ, 1851.

(10.) C. DARWIN.   A Monograph of the Sub-class Cirripedia, vol. 2, Balanidæ, Verrucidæ, &c.   1854.

(11.) MAX SCHULTZE.   Zoologische Skizzen, p. 189.   Zeitschrift für Wissenschaftl. Zoologie, vol. 4, 1853.

(12.) HESSE.   Mémoire sur les Métamorphoses que subissent pendant la période embryonnaire les Anatifs appelés Scalpels obliques.   Annales des Sciences Nat., vol. 11, 1859.

(13.) A. KROHN.   Beobachtungen über die Entwicklung der Cirripedien.   Archiv f. Naturgesch., vol. 25, 1859.

(14.) CLAPARÈDE.   Beobachtungen über Anatomie und Entwicklungsgeschichte der Wirbelloser Thiere.   Zur Entwicklung der Cirripedien.   1863.

(15.) A. PAGENSTECHER. Beiträge zur Anatomie und Entwicklungsgeschichte von Lepas pectinata. Zeit. f. Wiss. Zool., vol. 13, 1863.

(16.) F. MÜLLER. Für DARWIN. 1864.

(17.) F. DE FILIPPI. Ueber die Entwicklung von Dichelaspis darwinii. MOLESCHOTT, Untersuchungen, vol. 9, 1865.

(18.) METSCHNIKOFF. Ueber die Entwicklung von Balanus balanoides. Sitzungsb. der Versamml. Deutsch Naturf. zu Hannover, 1865.

(19.) A. GERSTAECKER. BRONN's Thierreich, vol. 5. Arthropoda. 1866.

(20.) C. CLAUS. Die Cypris-ähnliche Larva der Cirripedien und ihre Verwandlung in das festsitzende Thier. Marburg and Leipzig. 1869.

(21.) SPENCE BATE. The Impregnation of the Balani. Ann. Mag. Nat. Hist., Fourth Series, vol. 3, 1869.

(22.) MÜNTER and BUCHHOLZ. Ueber Balanus improvisus. Mittheilungen aus dem Naturwissenschaft. Vereine von Neu-Vorpommern und Rügen, vol. 1, 1869.

(23.) ANT. DOHRN. Untersuchungen über Bau und Entwicklung der Arthropoden. IX.—Eine neue Naupliusform (Archizoëa gigas). Zeitschrift für Wiss. Zoologie, vol. 20, 1870.

(24.) TARGIONI-TOZZETTI. Di una nuova specie di un nuovo genere di Cirripede di Lepadidæ. Bull. Soc. Ent. Ital., anno 4, pp. 84, 96.

(25.) C. E. HESSE. Description de la série complète des métamorphoses que subissent durant la période embryonnaire les Anatifes designés sous le nom de Scalpel oblique ou Scalpel vulgaire. Rev. Sciences Nat., Montpellier, 1874.

(26.) CARL BOVALLIUS. Om Balanidernas Utvickling. Embryologiska Studier. Stockholm. 1875.

(27.) A. GERSTAECKER. Ueber Ornitholepas Australis das Cypris-stadium einer Cirripedien Larve. Sitzb. d. Naturf. Ges. Berlin, 1875, pp. 113–115.

(28.) R. VON WILLEMOES-SUHM. On the development of Lepas fascicularis and the Archizoëa of Cirripedia. Phil. Trans., vol. 166, 1876.

(29.) C. CLAUS. Untersuchungen zur Erforschung der genealogischen Grundlage des Crustaceen-systems. Wien. 1876.

(30.) P. P. C. HOEK. Zur Entwicklungsgeschichte der Entomostraken. I. Embryologie von Balanus. Niederländisches Archiv für Zoologie, vol. 3, 1876-7.

(31.) ARNOLD LANG. Vorläufige Mittheilung über die Bildung des Stieles bei Lepas anatifera. Mittheilungen der Naturforschenden Gesellschaft in Bern. 1878.

(32.) ARNOLD LANG. Die Dotterfurchung von Balanus. Jenaische Zeitschrift, vol. 12, 1878.

(33.) ARNOLD LANG. Ueber die Metamorphose der Nauplius-Larven von Balanus. Mittheilungen der Aargauischen Naturforschenden Gesellschaft, vol. 1, 1863-1877.

(34.) F. M. BALFOUR. A Treatise on Comparative Embryology, vol. 1, 1880.

(35.) C. GROBBEN. Die Entwicklungsgeschichte von Cetochilus septentrionalis. Arbeiten a. d. Zoolog. Inst. z. Wien, vol. 3, 1881.

(36.) AGASSIZ, FAXON, and MARK. Selections from Embryological Monographs. I. Crustacea. FAXON.

(37.) R. SCHMIDTLEIN. Vergleichende Uebersicht über das Erscheinen grösserer pelagischer Thiere während der Jahre 1875-77. Mitth. Zool. Stat. Neapel., vol. 1, 1882.

(38.) LACAZE-DUTHIERS. Histoire de la Laura Gerardiae, type nouveau de Crustacé parasite. Paris. 1882.

(39.) P. P. C. HOEK. Report on the Cirripedia collected by H.M.S. Challenger. 1882-4.

(40.) N. NASONOV. Zur Embryonalen Entwicklung von Balanus. Zool. Anzeiger. 1885.

(41.) N. NASONOV. Izvyest. Moscow Univ., t. 52. (Development of Balanus improvisus.)

(42.) GILSON. La Cellule, tome 2. (Spermatogenesis.)

(43.) WEISMANN and ISHIKAWA. Weitere Untersuchungen zum Zahlengesetz der Richtungskörper. Zoologisches Jahrbuch, Morph. Abth., vol. 3, 1887.

(44.) M. NUSSBAUM. Vorläufiger Bericht. Sitzb. Akad. Berlin, 1887, p. 1051. (See also Abstract in Annals and Mag. of Nat. Hist., Series 6, vol. 1, 1888.)

(45.) P. P. C. HOEK. Tijdschrift Ned. Dierk. Vereeniging (2), vol. 3, Afl. 1, 1890, p. 33 (Development of Balanus).

(46.) SALVATORE LO BIANCO. Notizie biologiche riguardanti specialmente il periode di maturità sessuale degli animali del golfo di Napoli. Mitth. Zool. Stat. Neap., vol. 8, 1888.

(47.) R. KOEHLER. Recherches sur l'organisation des Cirripèdes. Archives de Biologie, vol. 9, 1889.

(48.) M. NUSSBAUM. Bildung und Anzahl der Richtungskörper bei Cirripedien. Zoolog. Anzeiger, vol. 12, 1890.

(49.) SOLGER. Die Richtungskörpechen von Balanus. Zoolog. Anzeiger, vol. 13, 1890.

(50.) GROOM and LOEB. Der Heliotropismus der Nauplien von Balanus perforatus und die periodischen Tiefenwanderungen pelagischer Tiere. Biologisches Centralblatt, vol. 10, 1890.

(51.) M. NUSSBAUM. Anatomische Studien an Californischen Cirripedien. Bonn. 1890.

(51A.) N. KNIPOVICHA. Materialui k poznaniyu gruppui Ascothoracida (with German abstract). St. Petersburg, 1892.

ANATOMY, &c., OF CIRRIPEDES.

(52.) R. WAGNER. Müller's Archiv für Anatomie und Physiologie. 1834.

(53.) H. MERTENS. Müller's Archiv für Anatomie und Physiologie. 1835.

(54.) MARTIN-SAINT ANGE. Mémoire sur l'Organisation des Cirripèdes et sur leurs rapports naturels avec les Animaux Articulés. 1835.

(55.) A. KROHN. Beobachtungen über das Cementapparat und die Weiblichen Zeugungsapparate einiger Cirripedien. WIEGMANN's Archiv für Naturgeschichte, vol. 25, 1859.

(56.) R. KOSSMANN. Arbeiten a. d. Zoolog.-zootom. Institut, Würzburg, vol. 1., 1874.

(57.) F. MÜLLER. Ueber Balanus armatus und einen Bastard dieser Art u. d. Balanus improvisus, var. assimilis Darw. Archiv für Naturgeschichte, 1867, vol. 1.

(58.) E. VAN BENEDEN. Développement des Sacculines. Bull. de l'Acad. Roy. de Belg., 1870.

COPEPODA AND PHYLLOPODA.

(59.) C. CLAUS. Zur Anatomie u. Entwicklungsgeschichte d. Copepoden. Archiv f. Naturgeschichte, vol. 24, 1858.

(60.) E. VAN BENEDEN. Recherches sur l'Embryogénie des Crustacés. IV. Anchorella, Lerneopoda, Branchiella, Hessia. Bull. de l'Acad. Roy. de Belgique, 2me Série, vol. 29, 1870.

(61.) C. CLAUS. Untersuchungen über die Organisation und Entwickelung von Branchipus und Artemia. Arbeiten. a. d. Zool. Inst. Wien, vol. 6, 1886.

(62.) C. CLAUS. Ueber d. Entwicklung, Organisation u. systematische Stellung d. Argulidæ. Zeit. f. Wiss. Zoologie., vol. 25, 1875.

(63.) F. VEJDOVSKY. Untersuchungen über d. Anat. u. Metaphor. v. Tracheliastes polycolpus. Zeit. f. Wiss. Zoologie., vol. 27, 1877.

(64.) P. P. C. HOEK. Zur Entwicklungsgeschichte der Entomostraken. II. Zur Embryologie der freilebenden Copepoden. Niederländisches Archiv für Zoologie, vol. 4, 1877-8.

(65.) C. GROBBEN. Zur Entwicklungsgeschichte d. Moina rectirostris. Arbeit. a. d. Zoologisch. Institute, Wien, vol. 2, 1879.

(66.) E. RAY LANKESTER. Observations and Reflections on the Appendages and on the Nervous System of Apus cancriformis. Quart. Journ. Microsc. Science, vol. 21, 1881.

(67.) F. URBANOWICZ. Beiträge zur Entwicklungsgeschichte der Copepoden. Kosmos, Lemberg, Jahrg. 10.

LEPTOSTRACA.

(68.) E. METSCHNIKOFF. Development of Nebalia (Russian), 1868.

(69.) C. CLAUS. Ueber den Organismus der Nebaliden und die systematische Stellung der Leptostraken. Arbeit a. d. Zoolog. Institute, Wien, vol. 8, 1889, p. 101.

SCHIZOPODA.

(70.) E. VAN BENEDEN. Recherches sur l'Embryogénie des Crustacés. II. Développement des Mysis. Bull. Acad. Roy. de Belgique, Second Series, vol. 28,
1869.

DECAPODA.

(71.) P. MAYER. Zur Entwicklungsgeschichte d. Dekapoden. Jenaische Zeitschrift,
vol. 11, 1877.
(72.) W. FAXON. On the Development of Palæmonetes vulgaris. Bull. of the Mus.
of Comp. Zool., Harvard, Cambridge, Mass., vol. 5, 1889.
(73.) W. REICHENBACH. Studien zur Entwicklungsgeschichte des Flusskrebses.
Abh. d. Senckenbergischen Naturforschenden Gesellschaft zu Frankfurt am
.   Main, vol. 14.
(74.) J. S. KINGSLEY. The Development of Crangon vulgaris. Second Paper.
Essex Inst. Bulletin (Salem, Mass.), vol. 18.

ARTHROSTRACA.

(75.) H. RATHKE. Zur Morphologie. Reisebemerkungen aus Taurien. Riga u.
Leipzig, 1837.
(76.) E. VAN BENEDEN. Recherches sur l'Embryogénie des Crustacés. I. Observations sur le Développement de l'Asellus aquaticus. Bull. de l'Acad. Roy. de
Belgique, 2me Série, vol. 28, 1869.

HEXAPODA.

(77.) A. KOROTNEFF. Die Embryologie der Gryllotalpa. Zeitschrift für Wissen-
schaftliche Zoologie, vol. 41, 1885.

GENERAL.

(78.) A. WEISMANN. Studies in the Theory of Descent. English Translation,
1882.
(79.) J. LOEB. Der Heliotropismus der Thiere. Würzburg, 1890.
(80.) A. LANG. Lehrbuch der Vergleichenden Anatomie. Jena, 1888.
(81.) MILNES MARSHALL. Opening Address to the Biological Section of the British
Association. Nature, 1890.
(82.) KORSCHELT and HEIDER. Lehrbuch der Vergleichenden Entwicklungsgeschichte der Wirbellosen Thiere. Specieller Theil, Zweites Heft, 1891.

DESCRIPTION OF PLATES.

EXPLANATION OF LETTERING.

abd.reg.	Abdominal region.	*frl.gl.s.*	Secretion of ditto.
an.arc.	Anal arc.	*frl.h.*	Fronto-lateral horn.
an.dil.	Dilator of anus.	*gn.*	Gnathobase of antennæ.
ant.[1]	Antennules.	*gn.m.*	Muscle to ditto.
ant.[2]	Antennæ.	*gr.mat.*	Granular matter.
anus.	Anus.	*gt.c.m.*	Circular muscles of stomach
app.	Appendage.		and intestine.
app.d.musc.	Dorsal muscles to appen-	*gt.l.m.*	Longitudinal muscles of
	dages.		ditto.
app.v.musc.	Ventral muscles to appen-	*int.*	Intestine (proctodæum).
	dages.	*lat.gl.*	Lateral gland.
arch.	Archenteron (stomach).	*lbr.*	Labrum.
ax.gl.	Axial gland of labrum.	*lbr.br.*	Bristles on labrum.
ax.gl.n.	Nucleus of ditto.	*lbr.c.*	Cells at sides of labrum.
ax.gl.fi.	Fibre to axial gland.	*lbr.dist.*	Distal lobe of labrum.
b.c.	Body cavity.	*lbr.n.*	Nerve to labrum.
bl.	Blastopore.	*lbr.prox.*	Proximal lobe of labrum.
br.	Brain (anterior lobes).	*mer.*[1], *mer.*[2], &c.	Merocytes.
br.acc.l.	Accessory lobes of brain.	*mes.c.*	Mesoderm cells.
br.c.l.	Central lobe of brain.	*mes.'*	Mesoblast of Nauplius.
br.p.l.	Posterior lobes of brain.	*mnd.*	Mandibles
car.	Carapace.	*mo.*	Mouth.
c.sp.	Caudal spine.	*mx.arc.*	Maxillary arc.
c.o.c.	Circum-œsophageal connec-	*m.pu.*	Male pro-nucleus.
	tives.	*Npl.eye.*	Nauplius eye.
c.t.	Connective tissue.	*œs.*	Œsophagus (stomodæum).
d.b.	Dorsal body.	*œs.d.m.*	Dilator muscles of ditto.
ect.	Ectoderm.	*o.g.*	Oil globules.
ect.n.	Nuclei of ectoderm.	*pmx.bd.*	Premaxillary band.
end.	Endoderm.	*ppm.*	Protoplasm of blastoderm.
ex.m.arc.	Extra-maxillary arc.	*proct.*	Intestine (probably = proc-
fl.arc.	Flexor arc.		todæum).
fl.ta.	Flexor of the tail.	*pv.*[1]	First polar body.
fr.fil.	Frontal filaments.	*pv.*[2]	Second polar body.
fr.fil.bs.	Base of ditto.	*st.*	Stomach.
frl.gl.	Fronto-lateral gland.	*st.gl.c.*	Glandular cells of stomach.
frl.gl.n.	Nucleus of ditto.	*st.b.*	Stellate bodies.

stom.	Œsophagus (probably = stomodæum).	*ves.t.*	Vesicular tissue.
sub-œs.g.	Sub-œsophageal ganglion.	*vt.m.*	Vitelline membrane.
ta.	Tail (thorax-abdomen).	*yk.*	Yolk.
ta.th.	Ectodermal thickening in thorax-abdomen (ventral plate).	*yk.end.*	Yolk-endoderm.
		yk.end.c.	Yolk-endoderm cells.
		yk.end.n.	Yolk-endoderm nuclei.
		yk.gr.	Yolk granules.
ta.sp.	Strong pair of spines on tail.	*yk.n.*	Nuclei of meso-hypoblast cells.
th.reg.	Thoracic region.		

PLATE 14.

LEPAS ANATIFERA. (Figs. 1–19. × 300.)

Stage A.

Fig. 1. Unfertilized ovum, just laid.

Fig. 2. Ovum just after giving off the first polar body (*pv*[1].).

Fig. 3. Ovum undergoing retractions within the newly-formed peri-vitelline membrane.

Fig. 4. Ovum just after giving off the second polar body (*pv*[2].).

Fig. 5. Ovum during the process of separation of the yolk and protoplasm, showing the commencing formation of the first blastomere and central mass of protoplasm.

Fig. 6. Ovum showing the first blastomere clearly defined peripherally by a sharp line, while, internally, it is still in connection with the central mass of protoplasm. No nucleus is visible without preparation.

Fig. 7. Somewhat later stage in which the segmentation-nucleus has become visible as a clear round spot. The first blastomere is a larger one than that of fig. 6.

Stage B.

Fig. 8. Ovum in which the first blastomere is now clearly marked off by a constriction, and in which the nucleus has divided. One of the daughter-nuclei is passing towards one side of the central mass of protoplasm, which is still in communication with the protoplasm of the first blastomere.

Fig. 9. Embryo in which the first blastomere has become definitely cut off from the yolk by a transverse wall. One of the daughter-nuclei of the segmentation nucleus has passed to the periphery of the yolk, where new protoplasm is forming at the expense of the latter, the central mass of protoplasm having likewise become peripheral. The protoplasm of the newly-forming second blastomere is seen to be still in connection with the yolk.

Fig. 10. Similar stage in which the basal plane was oblique.

Fig. 11. Similar stage in which the first blastomere is not marked off by a constriction. The nuclei are large and clear.

Fig. 12. Lateral view of a stage in which the second blastomere has become pretty clearly defined peripherally. Both blastomeres (separated by an oblique basal plane in this case) are dividing transversely.

Fig. 13. Stage in which the second blastomere is arising at some distance from the first, which, in the meantime, is dividing transversely.

Fig. 14. Similar stage, but the second blastomere is in contact with the first, and its nucleus lies in the plane of division of the latter.

Fig. 15. Stage, with three blastoderm cells, and a fourth appearing on the left-hand side of II.

Fig. 16. Stage, with three blastomeres, and a fourth, nearly completed, on the right-hand side of II.

Fig. 17. Stage, with three blastomeres, and a fourth emerging beneath II. Ventral view.

Fig. 17A. Stage, with three blastomeres, and two new ones arising by division of a merocyte before it emerges from the yolk. Ventral view.

Fig. 18. Stage, with six blastomeres, one terminal (anterior), two dorso-lateral, two ventro-lateral, and one approximately ventro-lateral (dividing).

Fig. 19. Stage with seven blastomeres, two dorso-lateral, four ventro-lateral, one ventral, and one emerging ventro-laterally.

\

PLATE 15.

LEPAS ANATIFERA. (Figs. 20–39. × 300.)

Fig. 20. Lateral view of stage with fifteen cells, one dorsal, four dorso-lateral, six lateral, and four ventral.

Fig. 21. Stage in which the yolk is largely covered by the blastoderm; blastopore still large. The covered part of the yolk is shaded.

Fig. 22. Further stage; blastopore quite small (the yolk seen to project through it near the posterior end on the right-hand side). The yolk is faintly shaded.

Fig. 23. Stage in which the blastopore is closed by the formation of a new blastoderm cell over the uncovered part of the yolk. The yolk (which is shaded) is still undivided.

Stage C.

Fig. 24. Stage in which the yolk (meso-hypoblast) has just divided by an oblique furrow.

Fig. 25. Stage showing the production of the mesoblast of the Nauplius from the hind part of the two meso-hypoblast cells.

Fig. 26. Stage with three endoderm cells.

Fig. 27. Stage with four endoderm cells.

Fig. 28. Side view of stage with six endoderm cells, showing commencing segmentation of the body into three regions.

Stage D.

Fig. 29. Ventral view of stage in which the body is clearly divided into three regions.

Fig. 30. Same stage, side view. The furrows are seen to die out on the sides of the body.

Fig. 31. Same stage, dorsal view.

Stage E.

Fig. 32. Stage in which the three pairs of appendages are first marked out. Ventral view. *Ant.*[1], antennule; *ant.*[2], antennæ; *mn.*, mandible; *ta.*, tail.

Fig. 33. Same stage, lateral view.

Fig. 34. Dorsal view of same stage, showing the free ends of the appendages meeting in the mid-dorsal line.

Stage F.

Fig. 35. Ventral view of stage in which the appendages are well developed, but still short.

Fig. 36. Same stage, dorsal view.

Fig. 37. Lateral view of stage in which the appendages are longer, and directed obliquely backwards and upwards. *Ant.*[1], antennule; *ant.*[2]. antennæ; *mn.*, mandible; *ta.*, tail; *lbr.*, labrum. The hinder end of the labrum is marked by a slight notch at the level of the letters *ant.*[1].

Fig. 38. Stage in which the appendages are still of moderate length : the setæ becoming visible at the tips of the appendages. Ventral view. Lettering as in fig. 37. The tip of the labrum is seen opposite the middle of the antennæ : the endoderm is shaded.

Fig. 39. Lateral view of stage in which the appendages are a little longer. The endoderm is faintly shaded.

PLATE 16.

LEPAS ANATIFERA.

Fig. 40. Same stage as in fig. 39 : dorsal view. The boundaries of the endoderm
cells are clearly seen. × 300.

Stage G.

Fig. 41. Dorsal view of a Nauplius approaching maturity, but still enclosed within
the peri-vitelline membrane. *Yk.end.*, yolk-endoderm ; *c.sp.*, caudal
spine ; *ta.*, tail ; *br.*, brain (the Nauplius eye is not yet visible); *frl.gl.*,
fronto-lateral gland ; *frl.h.*, fronto-lateral horns.

Stage H.

Fig. 42. Dorsal view of a Nauplius removed from the peri-vitelline membrane shortly
before hatching. *Arch.*, stomach ; *br.*, brain ; *c.sp.*, caudal spine, telescoped
within the body ; *frl.gl.*, fronto-lateral gland ; *frl.h.*, fronto-lateral horn ;
gr.mat., granular matter ; *npl.eye*, Nauplius eye.

LEPAS PECTINATA.

Fig. 43. Ovum of *Lepas pectinata* after formation of the second polar body (*pv²*).
Fig. 44. Ovum contracting during the process of separation of the protoplasm and
yolk. The protoplasm is commencing to collect at the anterior end.
Fig. 45. Same ovum, five minutes afterwards ; a good deal of the protoplasm is now
seen at the anterior end.
Fig. 46. Same ovum, thirty minutes later ; the segmentation-nucleus is now visible
as a faint clear spot.
Fig. 47. Later stage of another ovum ; the first blastomere marked off by a deep
constriction.
Fig. 48. Embryo in which the first blastomere was separated from the yolk by an
oblique basal plane ; the second blastomere forming from the yolk.

BALANUS PERFORATUS. (Figs. 49–59. × 265.)

Stage A.

Fig. 49. Ovum of *Balanus perforatus* in which the protoplasm has segregated from
the yolk.

2 F 2

Stage B.

Fig. 50. Stage in which the first blastomere has been cut off from the yolk, and a second is arising from the latter.

Fig. 51. Stage in which the second blastomere is emerging, and the first dividing transversely.

Fig. 52. A similar, but somewhat later stage, in which the second blastomere is well marked off peripherally from the yolk.

Fig. 53. Stage with three blastomeres, and a fourth and fifth arising by division of a single merocyte. Another example was similar, but the merocyte was undivided.

Fig. 54. Stage with three blastomeres, the unpaired one (II.) dividing transversely.

Fig. 55. Dorsal view of stage with three blastomeres, a new one arising near the posterior end.

Fig. 56. Ventral view of similar stage.

Fig. 57. Stage with the blastomeres, two of which (probably all three*) are dividing ; a merocyte dividing in the yolk.

Fig. 58. Stage with four blastomeres ; a new one arising near the posterior end.

Fig. 59. Stage with four blastomeres, one (ventral) dividing transversely; a merocyte emerging on the right-hand side.

PLATE 17.

BALANUS PERFORATUS. (Figs. 61–78. × 265.)

Fig. 61. Stage with five blastomeres, a sixth arising from the yolk.

Fig. 62. Stage with seven blastomeres, one terminal, one ventral, two ventro-lateral, two lateral, and one dorsal ; an eighth arising from the yolk: Ventral view. The Roman numerals refer to the successive merocytes.

Fig. 63. Similar stage, ventral blastomere dividing.

Figs. 64–66. Later stages, showing further growth of blastoderm over yolk.

Fig. 67. Stage in which the blastopore is an irregular space becoming further diminished in size by the appearance of a new merocyte.

Fig. 68. Stage in which the blastopore is still smaller ; a new blastomere is arising at the apex of the projecting portion of the yolk.

Fig. 69. Stage in which the blastopore is very small.

Fig. 70. Stage in which the blastoderm is just completed by a merocyte emerging at the apex of the yolk to form a low blastomere just anterior to the most posteriorly situated cell. The yolk is still undivided.

Fig. 71. Similar stage, in which the outline of the yolk is confluent with that of the cell which closed the blastopore. The yolk still undivided.

---

* I did not succeed in seeing the opposite side of this egg.

Stage C.

Fig. 72. Stage in which the yolk (meso-hypoblast) is divided in two.

Fig. 72a and 72b. Two lateral views of stages with four endoderm cells.

Fig. 73. Lateral view of stage with six endoderm cells.

Fig. 75. Ventral view of approximately same stage as that of fig. 73. The nuclei of the endoderm cells appear as dark stellate spots.

Stage F.

Fig. 76. Lateral view of stage with short appendages. $ant.^1$ antennules; $ant.^2$ antennæ; $mn.$, mandibles; $ta.$, tail.

Fig. 77. Dorsal view of stage in which appendages are rather short; the second and third pairs are bifid at their extremities. Lettering as in fig. 76.

Fig. 78. Same stage, side view. The nuclei of the endoderm cells may be sometimes seen dividing. Lettering as in fig. 76.

PLATE 18.

BALANUS PERFORATUS.[*] (Figs. 79-82. × 265.)

Fig. 79. Same stage as in fig. 78; ventral view.

Fig. 80. Stage in which the appendages are of moderate length; labrum ($lbr.$) well-marked; lettering as in fig. 82; ventro-lateral view.

Stage G.

Fig. 81. Lateral view of stage in which the appendages are long; Nauplius eye as yet absent; lettering as in fig. 82.

Stage H.

Fig. 82. Nauplius nearly ready to hatch; side view; $ant.^1$, antennule; $ant.^2$, antenna; $mn.$, mandible; $ta.$, tail; $c.sp.$, caudal spine; $npl.eye$, Nauplius eye (black); $yk.end.$, yolk-endoderm. At the side of the intestine and hind end of the stomach are seen flat granular cells.

CHTHAMALUS STELLATUS. (Figs. 83–96. × 360.)

Stage A.

Fig. 83. Ovum of *Chthamalus stellatus* in which the protoplasm and yolk are segregating. $pv.^2$, second polar body.

[*] The colours in figs. 79–82 should be as in figs. 61–78.

Fig. 84. Ovum in which segregation of the protoplasm and yolk is complete, the
boundary being transverse. The nucleus has appeared as a small clear
spot in the protoplasm, just anterior to the basal plane; around this the
protoplasm, preparatory to the division of the nucleus, has assumed a
radial arrangement; $pv.^{2}$, second polar body.

Fig. 85. Example of an egg in which the basal plane was oblique.

Stage B.

Fig. 86. Example of an egg in which the second blastomere is arising before the
separation of the first from the yolk along the basal plane has taken
place.

Fig. 87. Egg showing the second blastomere arising from the yolk; the first having
been cut off by an oblique basal plane.

Fig. 88. Stage showing second blastomere arising from the yolk while the first is
dividing.

Fig. 89. Stage with three blastomeres, and a fourth arising from the yolk on the left-
hand side of II. (the cell on the left side of the figure).

Fig. 90. Stage with three blastomeres, and a fourth arising from the yolk in the
middle line.

Fig. 91. Stage with five or six blastomeres.

Fig. 92. Stage in which the yolk is half covered by the blastoderm; a new blastomere
is emerging on the right-hand side of the yolk.

Fig. 93. Similar stage to that in fig. 92, but the merocyte in connection with the
nearest blastomere by a nuclear spindle, seen by focussing below the
surface.

Fig. 94. Stage in which the blastopore is being closed by the emergence of a
merocyte.

Stage C.

Fig. 95. Stage in which the yolk (meso-hypoblast) is divided in two, each half being
a cell with a nucleus and protoplasmic mass at its hinder end.

Fig. 96. Example in which the yolk had divided before the closure of the blastopore.

PLATE 19.

CHTHAMALUS STELLATUS.

Fig. 97. Stage showing commencement of the formation of the mesoblast of the
Nauplius; the yolk has lost its definite contour in the region where this

is taking place, and the mesoblastic cells themselves form a rather opaque mass (on the right-hand side of the figure). × 360.

Stage F.

Fig. 98. Dorsal view of embryo-Nauplius with rather short appendages, taken out of the egg-shell ; $ant.^1$, antennule ; $ant.^2$, antenna ; $mn.$, mandible. The yolk endoderm is shaded. × 360.

Stage H.

Fig. 99. Nauplius almost ready to hatch ; $ant.^1$, antennule ; $ant.^2$, antenna ; $mn.$, mandible ; $npl.eye$, Nauplius eye ; $lbr.$, labrum ; $ta.$, tail ; $c.sp.$, caudal spine.

LEPAS ANATIFERA. ·(Figs. 101–113. × 420.)

Stage A.

Fig. 100. Longitudinal section through ovum of *Lepas anatifera*, showing the nucleus dividing to form the second polar body ; $pv.^1$, first polar body ; the oil globules appear as clear round spaces.

Fig. 101. Oblique section of ovum in which the protoplasm and yolk are segregated, showing division of the segmentation-nucleus. This stage would be a little later than that shown in fig. 7.

Fig. 102. Oblique section passing through stage between those shown in figs. 8 and 9. The first blastomere has quite recently been cut off, and the spindle-fibres are still visible.

Fig. 103. Similar section of an egg in which the basal plane was oblique.

Fig. 104. A more longitudinal section of a rather later stage in which the two nuclei are completely disconnected ; *cf.* stage shown in woodcut $3f$ (p. 196).

Fig. 105. Longitudinal section of an egg at the stage shown in woodcut $3g$; *cf.* also figs. 9–11.

Stage B.

Fig. 106. Oblique section of an egg at stage a little later than that shown in fig. 17 ; *cf.* also fig. 17A ; a nucleus has recently been given off into the yolk from the third blastomere with which it is still in connection.

Stage C.

Figs. 107 and 108. Oblique sections passing through the two yolk cells (meso-hypoblast) which have each cut off a mesoblastic cell (*mes.*[1]); *cf.* wood-cut 3*q* (p. 197).

Fig. 109. Sections of slightly later stage, in which the mesoblast cell has divided.

Fig. 109*a.* Next one in the series of sections of the same embryo.

Fig. 110. Section of a still later stage in which the yolk-cells have cut off two more mesoblastic cells.

Fig. 111. Longitudinal section, showing yolk-endoderm divided into two; mesoblast cells behind. Lettering as in fig. 112.

Fig. 112. Longitudinal horizontal section of embryo at stage shown in fig. 28; several yolk-endoderm cells present (four intersected). *Ect.*, ectoderm; *mes.*[1], mesoderm; *yk. end.*, yolk-endoderm cell; *yk. end. n.*, nucleus of yolk-endoderm cell.

Stage D.

Fig. 113. Longitudinal vertical nearly median section of embryo at stage shown in figs. 29–31, in which the body is apparently segmented into three portions; the section shows the dorsal mesoblastic plate (*mes.*[1]).

PLATE 20.

LEPAS ANATIFERA. (Figs. 114–121*f.* × 420; figs. 122*a*–122*e.* × 375.)

Fig. 114. Transverse section of same stage as that shown in fig. 113.

Stage E.

Fig. 115. Transverse section through an embryo with short appendages (figs. 35 and 36), taken between two pairs of appendages.

Fig. 116. Transverse section of same stage, taken through a pair of appendages.

Stage F.

Fig. 117. Longitudinal horizontal section through an embryo of Stage F (fig. 37); the œsophagus (*stom.*) is cut transversely; the intestine (*proct.*), longitudinally; *app.*, appendages; *ta.*, tail.

Fig. 118. Transverse section through anterior region of embryo of same stage showing the thin dorsal ectoderm, no longer covered by the free ends of the

antennules, the lower parts of which, with their mesoblastic tissue, are still closely adpressed to and not to be distinguished from the sides of the carapace ; ventral to the œsophagus is the tissue of the base of the labrum not clearly marked off in this section.

Fig. 119. Transverse section of embryo at same stage, taken through the *labrum* (*lbr.*) ; *app.*, appendage.

Fig. 120. Transverse section of embryo at same stage, taken through the intestine (*proct.*) ; *ect.*, ectoderm ; *app.*, appendage.

Stage G.

Fig. 121. Nearly median longitudinal vertical section of embryo-Nauplius with long appendages. *b.c.*, body cavity ; *br.*, brain ; *c.sp.*, caudal spine ; *ect.*, ectoderm ; *proct.*, intestine ; *sub.œs.g.*, sub-œsophageal ganglion ; *ta.*, tail ; *yk.end.*, yolk-endoderm ; *yk.end.n.*, yolk-endoderm nucleus. The boundaries between the individual yolk-endoderm cells are not clearly seen in this section.

Fig. 121a. Transverse section through embryo of same stage, taken anteriorly to the point at which the caudal spine separates from the tail (about the level of letters *ect.* in fig. 121). On each side of the tail are seen the exopodite (dorsal), and the endopodite (ventral) of the mandibles ; outside these come the exopodite (dorsal) and the endopodite (ventral) of the antennæ ; *b.c.*, body cavity.

Fig. 121b. Transverse section of embryo of same stage, taken some distance in front of the mouth. *lbr.*, labrum ; *œs.*, œsophagus ; *œs.c.m.*, circular muscles of œsophagus ; *yk.end.*, yolk-endoderm ; *yk.end.n.*, nuclei of yolk-endoderm cells.

Fig. 121c. Transverse section of same embryo, taken just anterior to mouth. Letters as in figs. 121b. *b.c.*, body cavity ; *c.o.c.*, circum-œsophageal connectives.

Fig. 121d. Transverse section of embryo of same stage, taken behind the mouth. Letters as in figs. 121b and 121c. *sub-œs.g.*, sub-œsophageal ganglion.

Fig. 121e. Half of a transverse section of embryo of same stage, passing through the fronto-lateral glands (*frl.gl.*). *frl.gl.n.*, nuclei of fronto-lateral glands.

Fig. 121f. Portions of longitudinal horizontal sections of embryo of same stage.

Fig. 122a–e. Sections showing the excavation of the yolk-endoderm to form the stomach, and the union of the œsophagus (*stom.*) and intestine (*proct.*) with the stomach. The boundaries of the yolk-endoderm cells are clearly seen ; the nuclei appear deeply stained. Figs. 122a and 122d pass through the stomach and œsophagus ; fig. 122d through the stomach and intestine ; and figs 122b and 122e through the stomach only.

BALANUS PERFORATUS.

Stage B.

Fig. 123. Nearly longitudinal section through an egg of *Balanus perforatus*, in which the first blastomere has been cut off from the yolk, and the second is forming (*mer.*). × 420.

PLATE 21.

BALANUS PERFORATUS. (Figs. 125–136*d*. × 420.)

Stage B.

Figs. 124*a* and *b*. Longitudinal sections in planes at right angles to one another, showing the first blastomere dividing transversely, and the second giving off a nucleus into the yolk. × 375.

Fig. 125. Nearly longitudinal section through an egg, with three blastomeres and a merocyte (*mer.*) in the yolk.

Fig. 126. Longitudinal section through an egg, with six blastomeres, and a seventh arising from the yolk.

Fig. 127. Section through a stage in which the blastopore is just being closed by a merocyte (*mer.*).

Stage C.

Fig. 128. Longitudinal section of an embryo in which the blastoderm is completed ; a single merocyte, with its nucleus (*mer.n.*), is seen in the yolk.

Fig. 129. Oblique section of a stage in which the yolk (*meso-hypoblast*) has just divided.

Fig. 130. Oblique section of stage with two yolk-cells (*meso-hypoblast*).

Fig. 131. Obliquely longitudinal section of stage, with about five or six endoderm cells. *mes.*[1], mesoblast.

Fig. 132. Transverse section, near hind end, of embryo at same stage. *mes.*[1], mesoblast.

Stage E.

Fig. 133. Transverse section of embryo with short appendages, before the appearance of the œsophagus, showing mid-dorsal groove.

Stage F.

Fig. 134. Transverse section through an embryo-Nauplius at Stage F, taken through the œsophagus (*stom.*). The labrum (*lbr.*) is seen as a low ventral projection ; *yk.end.n.*, nuclei of yolk-endoderm cells.

Fig. 135. Transverse section of the same embryo passing through the intestine (*proct.*), and appendages (*app.*).

Nauplius. Stage II.

Fig. 136a. Longitudinal section (considerably inclined to the sagittal plane) of Nauplius after the first moult, passing to one side of the labrum, mouth, and œsophagus. The communication between the stomach (*st.*) and intestine (*int.*) is well seen. The section also passes close to the anus. The "ventral plate" is cut parallel to its surface, so that its composition out of a single layer of cells is not clearly seen.

Fig. 136b. Longitudinal vertical section through posterior half of a Nauplius of Stage II., traversing stomach (*st.*), intestine (*int.*), sub-œsophageal ganglion (*sub-œs.g.*), and "ventral plate" (*ta.th.*).

Fig. 136c. Nearly transverse section, through Nauplius of same stage, passing through stomach (*st.*), œsophagus (*œs.*), circum-œsophageal connectives (*c.o.c.*), and labrum (*lbr.*). The œsophagus is cut through as it bends back dorsally and ventrally.

Fig. 136d. Nearly transverse section of Nauplius at same stage, passing through stomach (*st.*), mouth (*mo.*, on one side), labrum (*lbr.*), circum-œsophageal connectives (*c.o.c.*).

PLATE 22.

BALANUS PERFORATUS.

Nauplius. Stage II. (Figs. 137a–139b. × 420.)

Fig. 137a–d. Sections of Nauplius taken parallel to the labrum. *Ant.*[1], antennule; *ant.*[2], antenna; *app.d.m.*, dorsal muscles of appendages; *app.v.m.*, ventral muscles of appendages; *br.*, brain; *br.acc.l.*, accessory lobes of brain; *br.c.l.*, central lobe of brain; *br.p.l.*, posterior lobes of brain; *fr.fil.bs.*, base of frontal filament; *mnd.*, mandible; *npl.eye.*, Nauplius eye; *œs.*, œsophagus; *sub.œs.g.*, sub-œsophageal ganglion; *st.*, stomach.

Fig. 137a. Section passing immediately above mouth.

Fig. 137b. Section next but one higher.

Fig. 137c. Next section higher.

Fig. 137d. Next section higher.

Fig. 138. Horizontal section through anterior part of Nauplius, traversing brain, Nauplius eye, frontal filaments, stomach, and fronto-lateral glands. Letters as in figs. 137a–d.

Fig. 138a. Muscles as seen in section of same Nauplius.

Fig. 139a. Tranverse section of Nauplius taken in front of mouth, and just behind the U-shaped bend of the œsophagus. *ax.gl.*, axial gland of labrum ; *ax.gl.n.*, nucleus of axial gland ; *car.*, edge of carapace ; *c.o.c.*, circum-œsophageal connectives ; *gn.*, gnathobase of antenna ; *gn.m.*, muscle of gnathobase ; *œs.*, œsophagus ; *st.*, stomach.

Fig. 139*b*. Transverse section of same Nauplius passing through the intestine (*int.*) and " ventral plate " (*ta.th*) ; *car.*, carapace.

Fig. 139*c*. Next section but one behind 139*b*.

Stage I.

Fig. 140. Dorsal view of Nauplius after moulting once. *frl.h.*, fronto-lateral horns ; *gr.*, granular matter ; other letters as in figs. 137 and 139. × 220.

PLATE 23.

BALANUS PERFORATUS.

Stage II.

Fig. 141. Dorsal view of Nauplius after moulting once. *an.dil.*, dilator of anus ; *br.*, brain ; *br.acc.l.*, accessory lobes of brain ; *ect.*, ectoderm ; *fl.ta.*, flexor of the tail ; *fr.fil.*, frontal filament ; *fr.fil.bs.*, base of frontal filament ; *frl.gl.*, fronto-lateral gland ; *frl.gl.s.*, secretion of fronto-lateral glands ; · *frl.h.*, fronto-lateral horn ; *int.*, intestine with its circular muscles ; *lat.gl.*, lateral gland ; *ves.t.*, vesicular tissue. × 220.

Fig. 142. Ventral view of Nauplius of same stage. *ax.gl.*, axial gland of labrum ; *c.sp.*, caudal spine ; *œs.*, œsophagus ; *gn.*, gnathobase of antenna ; *gn.m.*, muscle to ditto ; *ta.*, tail ; *ta.sp.*, tail spines : other letters as in fig. 141. On the ventral surface is seen the "ventral plate" and. a number of rows of setæ. The longitudinal muscles of the gut are faintly seen through the setose region. × 220.

Stage I.

Fig. 143. Dorsal view of anterior half of a Nauplius of Stage I., in which are shown the brain, with the Nauplius-eye (*npl.eye*), and accessory lobes (*br.acc.l.*) ; the circum-œsophageal connectives (*c.o.c.*), and sub-œsophageal ganglion (*sub-œs.g.*) ; behind the eye is seen the "dorsal body" (*d.b.*), and behind this is seen the outline of the stomach (*arch.*) some of the cells of which are drawn : the optical section of the vertical part of the œsophagus is seen as a faint circle in the midst of these.

PLATE 24.

BALANUS PERFORATUS.

Nauplius, Stage II.

Fig. 144. Ventral view of the setose region of a Nauplius of Stage II., showing the arrangement of the bands of setæ.

Fig. 145. Ventral view of labrum of a Nauplius of Stage II. × 420. *ax.gl.*, axial gland of labrum; *ax.gl.fi.*, fibre running to ditto; *br.*, brain; *ect.*, ectoderm; *fr.fil.*, frontal filament; *fr.fil.bs.*, base of ditto; *lbr.c.*, cells at side of labrum; *npl.eye*, Nauplius eye.

Fig. 146. Ventral view of thoracic region of a Nauplius of Stage II. *arch.*, stomach; *fl.ta.*, flexor of tail; *gt.l.m.*, longitudinal muscles of stomach and intestine; *ta.sp.*, tail-spines; *ta.th.*, ectodermal thickening of tail, or "ventral plate."

Fig. 147. Similar view showing a more advanced condition of the ventral plate. × 420.

Fig. 148. Portion of carapace of Nauplius (Stage II.), showing excreted granules of methyl-blue. × 420.

Fig. 148a. Labrum of Nauplius (Stage II.), showing excreted granules of methyl blue. × 420.

CHTHAMALUS STELLATUS.

Fig. 149. Dorsal view of Nauplius (Stage II.) of *Chthamalus stellatus. app. d. musc.*, dorsal muscles of appendages; *br.*, brain; *br.acc.l.*, accessory lobes of brain; *int.*, intestine; *fr.fil.bs.*, base of frontal filaments; *frl.gl.*, fronto-lateral gland; *frl.gl.s.*, secreted spherules of ditto; *frl.h.*, fronto-lateral horns; *lat.gl.*, lateral gland; *st.*, stomach; *ves.t.*, vesicular tissue. The sub-cuticular network is shown on the right-hand side, and the deeper structures only on the left. × 212.

PLATE 25.

CHTHAMALUS STELLATUS.

Nauplius, Stage II.

Fig. 150. Ventral view of Nauplius (Stage II.) of *Chthamalus stellatus. ax.gl.*, axial gland of labrum; *br.*, brain, with Nauplius eye resting upon it; *c.sp.*, caudal spine; *fr.fil.* frontal filament; *lat.gl.*, lateral gland; *lbr.dist.*

distal lobe of labrum ; *lbr.prox.*, proximal lobe of ditto ; *ta.sp.*, tail-spine.
× 216.

Fig. 151. Lateral view of Nauplius (Stage II.) of *Chthamalus stellatus. anus; an.
arc*, anal arc ; *app.d.musc.*, dorsal muscles of appendages ; *lbr.*, labrum ;
st., stomach ; *ta.*, tail ; other letters as in fig. 150. × 220.

Fig. 152. View of portion of ventral surface of Nauplius (Stage II.) of *Chthamalus
stellatus. an.arc*, anal arc ; *c.sp.*, caudal spine ; *ex.mx.arc*, extra-maxillary arc ; *fl.arc*, flexor arc ; *fl.ta.*, flexor of tail ; *lbr.*, labrum ; *pmx.bd.*,
pre-maxillary band ; *ta.*, tail ; *ta.sp.*, tail-spine. × 265.

LEPAS ANATIFERA.

Stage I.

Fig. 153. Ventral view of a Nauplius of *Lepas anatifera* just hatched, showing *brain*
(*br.*), with the *frontal filaments* (*fr.fil.*) underneath the cuticle, and their
bases (*fr.fil.bs.*) in the brain on each side of the Nauplius eye ; the
labrum (*lbr.dist.* and *lbr.prox.*) ; the fronto-lateral horns (*frl.h.*) are bent
back parallel to the body, and the caudal spine and tail telescoped within
the body, their tips only being as yet external ; portions of the fronto-
lateral glands are seen just in front of the antennules, and a portion of
the setose region behind the labrum. × 220.

Fig. 153a. Dorsal view of same. *app.d.musc.*, dorsal muscles of appendages ; *arch.*,
stomach ; *c.sp.*, caudal spine ; *fl.ta.*, flexor of tail ; *frl.gl.*, fronto-lateral
gland ; *gr.mat.*, granular matter ; above the Nauplius eye is seen the
" dorsal body." × 220.

Stage II.

Fig. 154. Nauplius of *Lepas anatifera* just after the first moult, with the labrum
turned a little forwards, so as to expose the mouth (*mo.*), and the origin
of the ventral muscles running to the appendages (*app.v.musc.*), as well as
the sub-œsophageal ganglion (*sub.œs.g.*). The tail and caudal spine are still
telescoped within the body. × 140.

Fig. 155. Ventral view of a more advanced Nauplius of Stage II., not long after the
first moult, with the labrum (*lbr.*) turned forwards. The tail and caudal
spine are still telescoped within the body. × 220.

PLATE 26.

LEPAS ANATIFERA.

Stage II.

Fig. 156. Dorsal view of a fully-developed Nauplius (Stage II.) of *Lepas anatifera*.
an.dil., dilator of anus; *app.d.musc.*, dorsal muscles of appendages; *br.*,
brain; *d.b.*, dorsal body; *fl.ta.*, flexor of tail; *frl.h.*, fronto-lateral horn;
frl.gl., fronto-lateral gland; *frl.gl.n.*, nuclei of ditto; *st.*, stomach, followed by intestine with its circular muscles; *st.b.*, stellate body; *ves.t.*,
vesicular tissue; in front of the anterior margin are seen the frontal
filaments. × 220.

PLATE 27.

LEPAS ANATIFERA.

Nauplius. Stage II.

Fig. 157. Ventral view of a fully-developed Nauplius after moulting *once* (Stage II.).
an.arc, anal arc; *ant.arc*, anterior arc; *ax.gl.*, axial gland of labrum;
c.sp., caudal spine; *d.b.*, dorsal body; *exm.arc*, extra-maxillary arc;
fl.arc, flexor arc; *frl.gl.*, fronto-lateral gland; *frl.gl.s.*, secreted spherules
of ditto; *lbr.*, labrum; *œs*, œsophagus; *ta.*, tail; *ta.sp.*, tail-spine. × 220.

Fig. 158. Ventral view of labrum and part of nervous system of a Nauplius of the
same stage. *Ax.gl.*, axial gland of labrum; *br.*, brain; *c.o.c.*, circum-
œsophageal connectives; *lbr.c.*, cells at sides of labrum; *lbr.dist.*, distal
lobe of labrum; *lbr.prox.*, proximal lobe of labrum; *œs.*, œsophagus with
its circular muscles. × 420.

Fig. 159. View of labrum of Nauplius of same stage turned forwards to expose the
mouth and hinder part of the nervous system. *Mo.*, mouth; *c.o.c.*, circum-
œsophageal connectives; *sub-œs.g.*, sub-œsophageal ganglion: other letters
as in fig. 158.

Fig. 160. View from below of setose region of Nauplius of same stage. *Fl.ta.*, flexor
of tail; *prmx.bd.*, premaxillary band; *ta.th.*, initial stage of ventral
plate: other letters as in fig. 157.

LEPAS PECTINATA.

Stage I.

Fig. 161. Nauplius of *Lepas pectinata* just hatched. *App.d.m.*, dorsal muscles to
appendages; *br.*, brain; *frl.gl.*, fronto-lateral glands at base of fronto-
lateral horns; *gr.mat.*, granular matter; *int.*, intestine; *st.*, stomach.
× 220.

PLATE 28.

LEPAS PECTINATA.

Nauplius. Stage II.

Fig. 163. Ventral view of setose region of Nauplius (Stage II.) of *Lepas pectinata.*
    × 420.

CONCHODERMA VIRGATA.

Nauplius. Stage I.

Fig. 164. Dorsal view of Nauplius of *Conchoderma virgata* just hatched ; *app.d.m.*,
    dorsal muscles of appendages ; *br.*, brain, with Nauplius eye ; *c.sp.*, caudal
    spine, telescoped within the body, together with the tail ; *frl.h.*, fronto-
    lateral horn ; *frl.gl.*, fronto-lateral gland ; *gr.mat.*, granular matter ; *st.*,
    stomach. × 220.
Fig. 165. Dorsal view of Nauplius of *Conchoderma virgata* some time after the first
    moult. The tail, caudal spine, and setæ are beginning to evaginate ;
    fl.ta., flexor of tail ; *frl.fil.*, frontal filaments ; *frl.gl.s.*, secretion of
    fronto-lateral gland ; *ta.*, tail ; other letters as in fig. 164. × 220.
Fig. 166. Ventral view of more advanced condition of Nauplius of same Stage (II.),
    showing the further evagination of the tail and caudal spine ; *br.*, brain ;
    c.sp., caudal spine ; *lbr.dist.*, distal lobe of labrum ; *lbr.prox.*, proximal
    lobe of labrum ; *œs.*, œsophagus ; *ta.*, tail. × 220.

DICHELASPIS DARWINII.

Nauplius. Stage I.

Fig. 167. Dorsal view of Nauplius of *Dichelaspis Darwinii* in the first stage ; *arch.*,
    stomach ; *ax.gl.*, axial gland of labrum ; *br.*, brain ; *fl.ta.*, flexor of tail ;
    lbr.dist., distal lobe of labrum ; *lbr. prox.*, proximal lobe of labrum ;
    npl.eye, Nauplius eye ; *œs.*, œsophagus. The tail and caudal spine are
    telescoped within the body, and the fronto-lateral horns are seen lying
    parallel to the sides of the body. × 220.
Fig. 168. Ventral view of a Nauplius of *Dichelaspis Darwinii* at Stage II, as made
    out from a number of mounted imperfect specimens ; *ax.gl.*, axial gland of
    labrum ; *c.sp.*, caudal spine ; *fl.ta.*, flexor of tail ; *fr.fil.*, frontal filaments ;
    frl.h., fronto-lateral horn ; *npl.eye*, Nauplius eye ; *ta.*, tail ; *ta.th.*, rudiment
    of ventral plate. × 110.

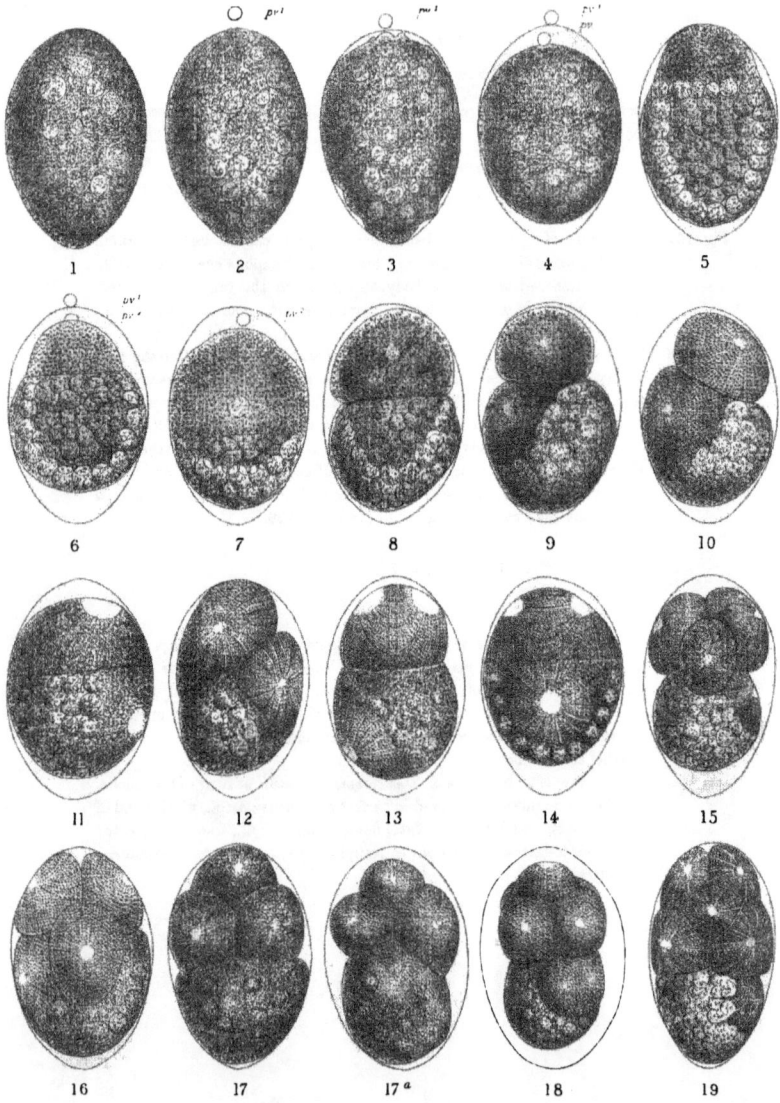

Phil. Trans. 1894 B *Plate* 1

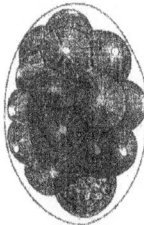

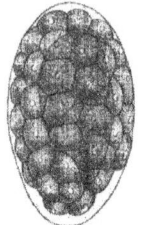

20 21 22 23 24

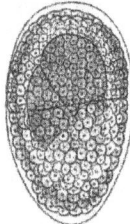
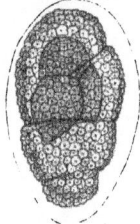

25 26 27 28 29

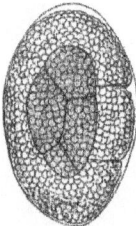
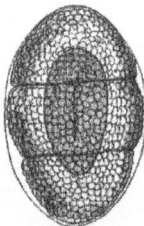
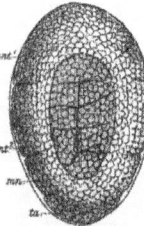
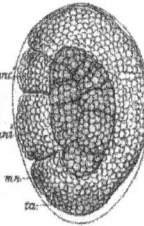
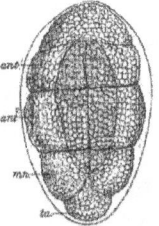

30 31 32 33 34

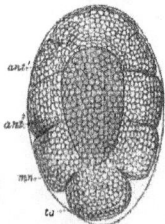

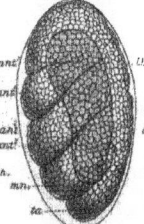
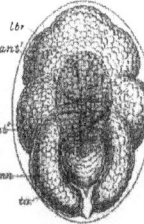
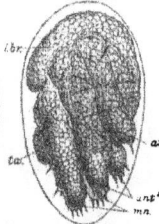

35 36 37 38 39

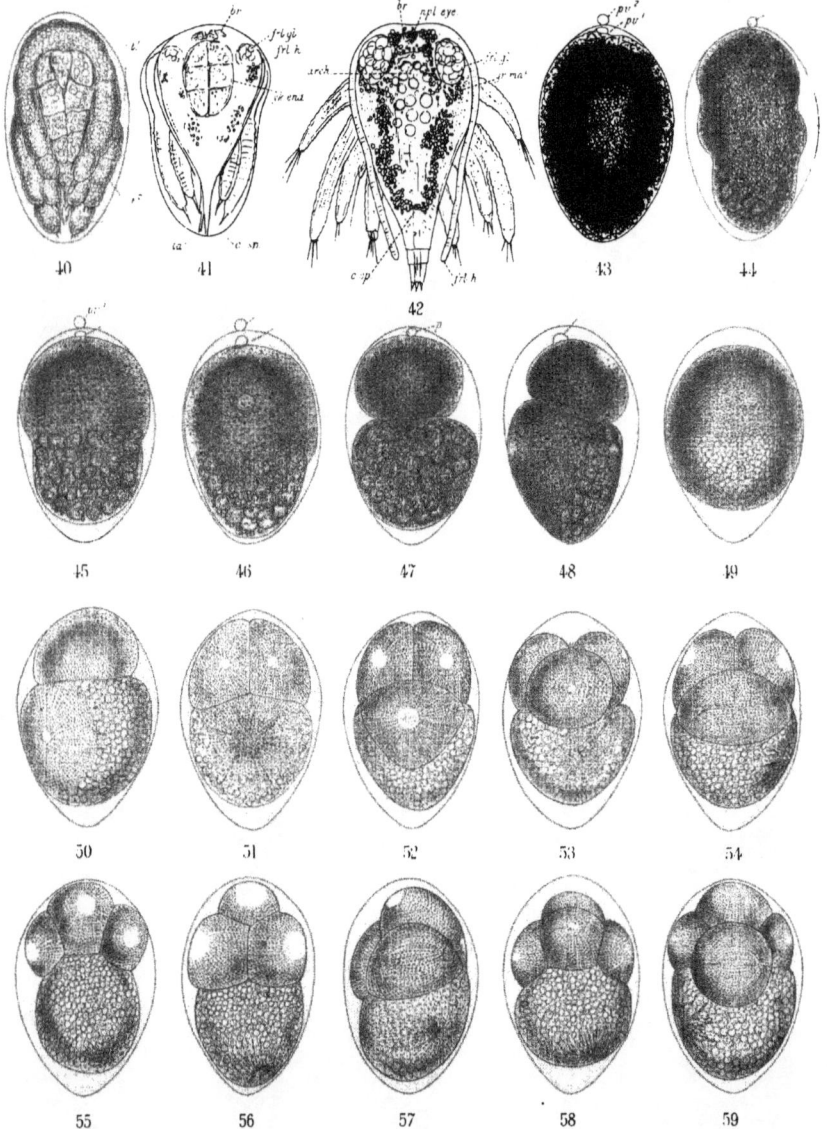

40   41   42   43   44

45   46   47   48   49

50   51   52   53   54

55   56   57   58   59

(Figs 40-42 , Lepas anatifera ; Figs 43-48 , Lepas pectinata ; Figs 49-59 , Balanus perforatus)

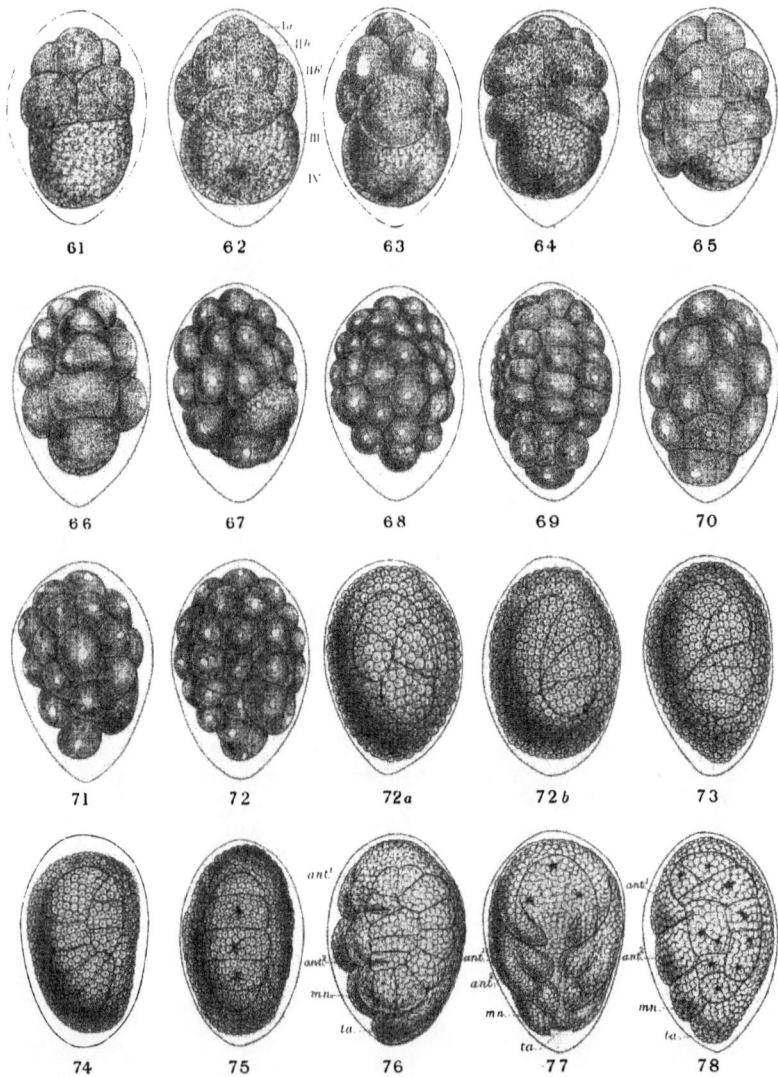

(Fig.ˢ 61 - 78 , Balanus perforatus.)

(Fig⁵ 79-82 , Balanus perforatus ; Fig⁵ 83-96 , Chthamalus stellatus)

(Figs 97-99, Chthamalus stellatus ; Figs 100-113, Lepas anatifera.)

Phil. Trans. 1894 B. Plate 20.

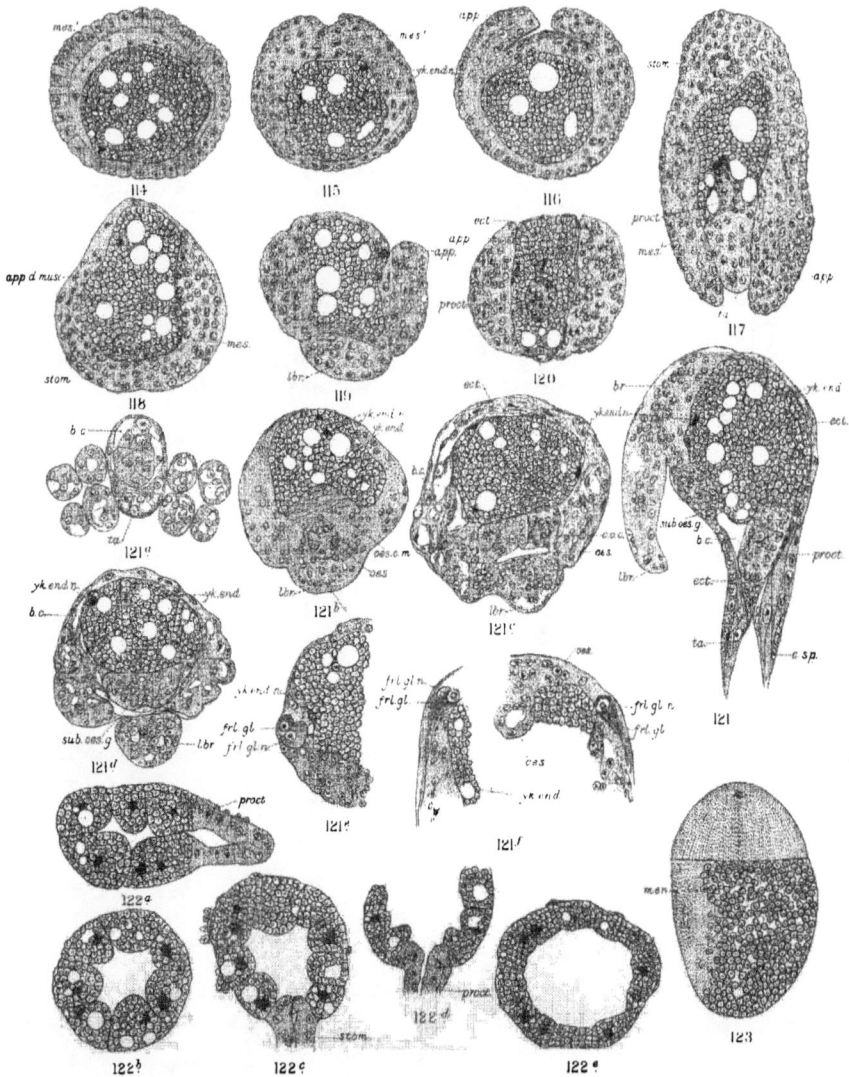

(Figs 114–122, Lepas anatifera ; Fig. 123, Balanus perforatus.)

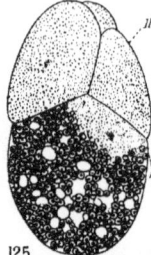

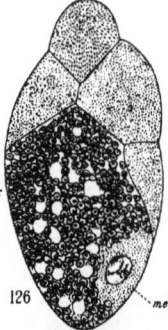

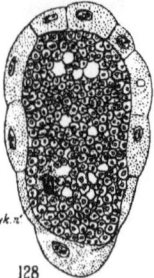

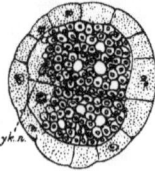

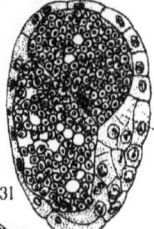

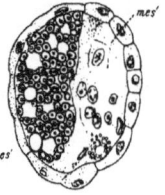

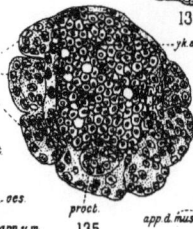

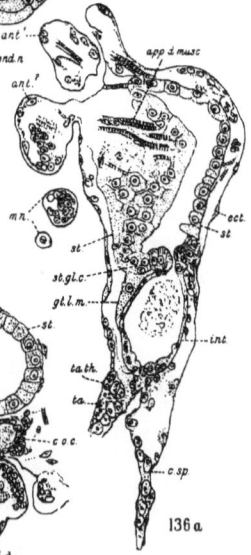

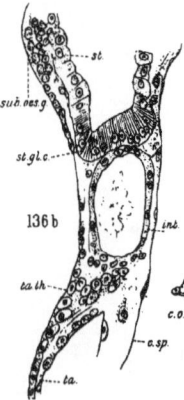

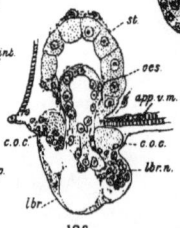

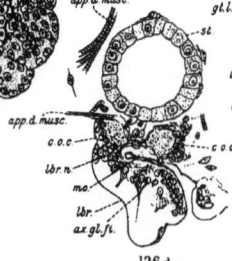

137a

137b

138

137c

137d

138a

139a

139b

139c

140

(Figs 137-140. Balanus perforatus.)

Groom.

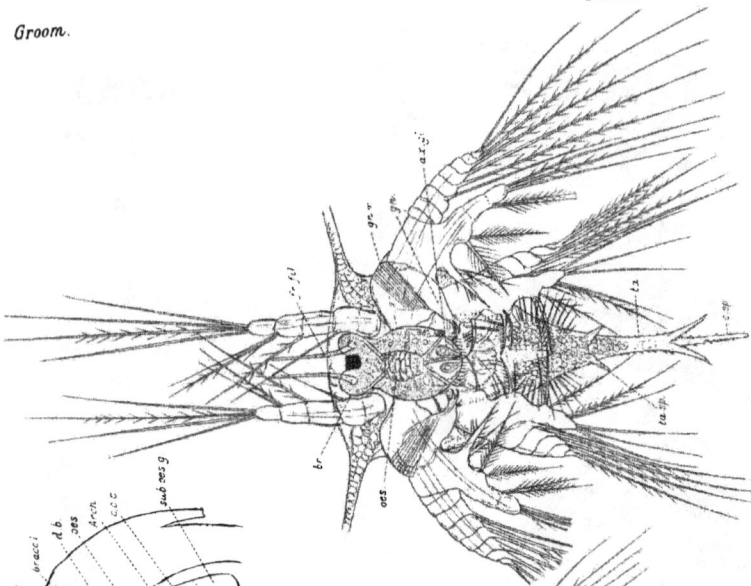

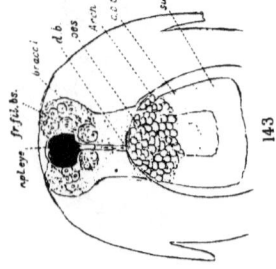

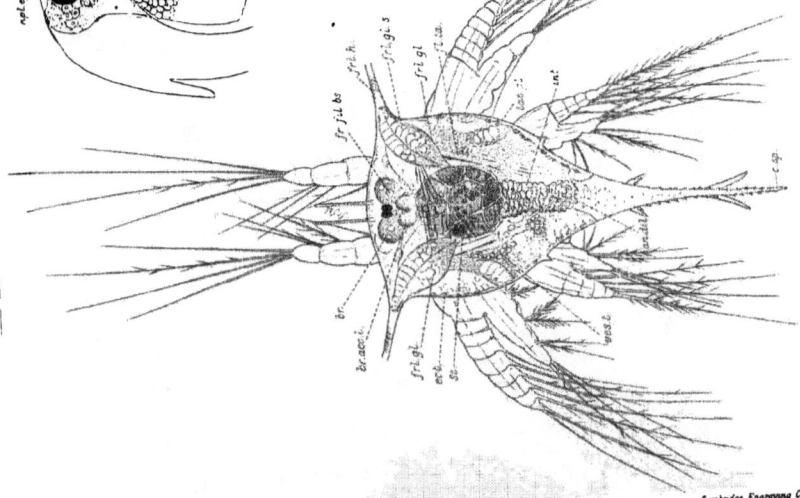

(Figs 141–143. Balanus perforatus.)

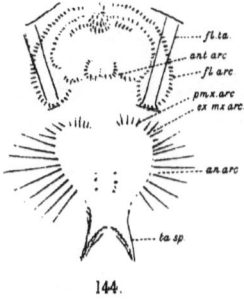

144.

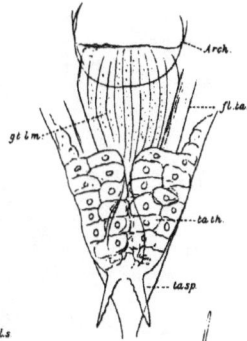

146.

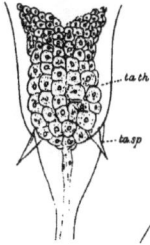

147.

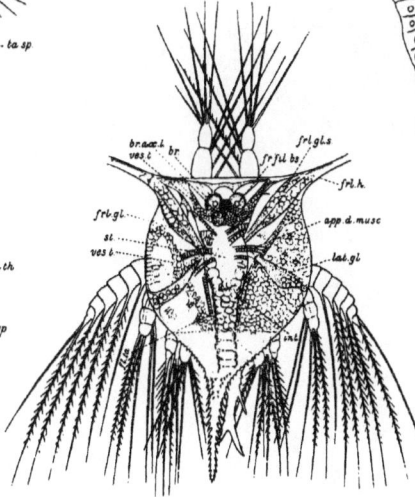

149.

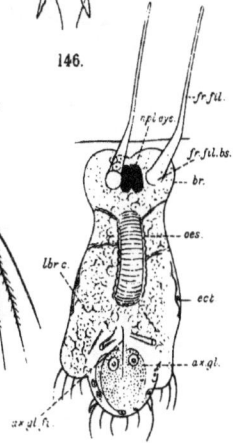

145.

148

148a.

(Fig.^{s} 144-148, Balanus perforatus.   Fig. 149, Chthamalus stellatus)

Del. Theo T Groom

Cambridge Engraving Company.

Phil. Trans. 1894 B. Plate 25.

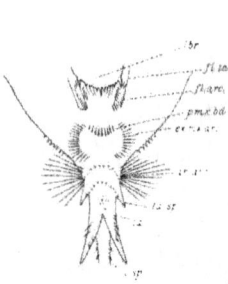

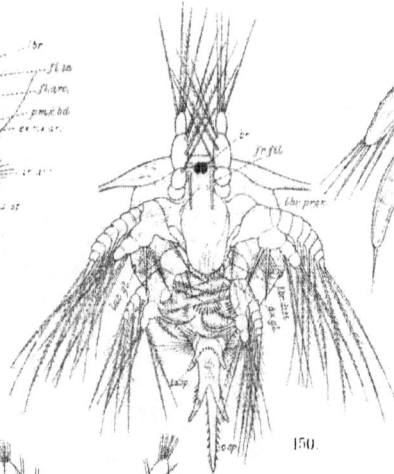

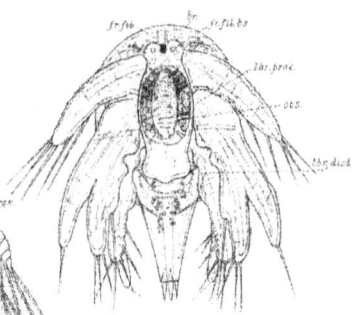

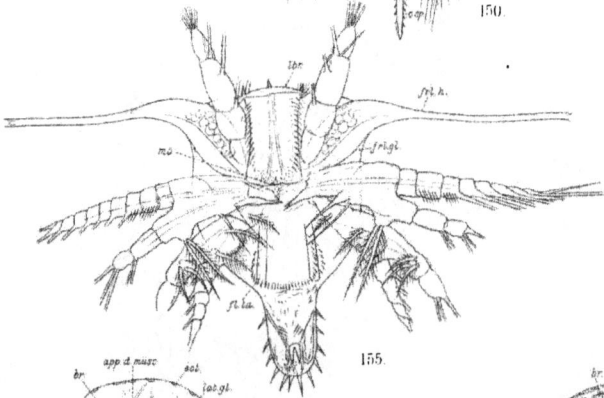

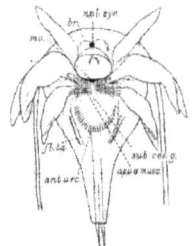

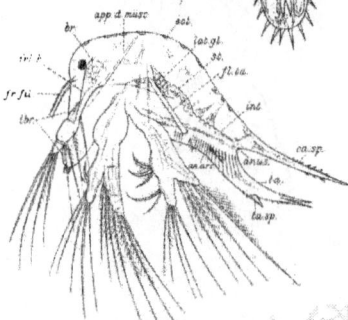

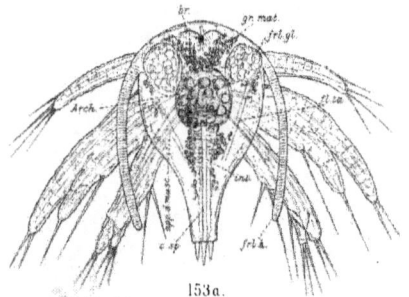

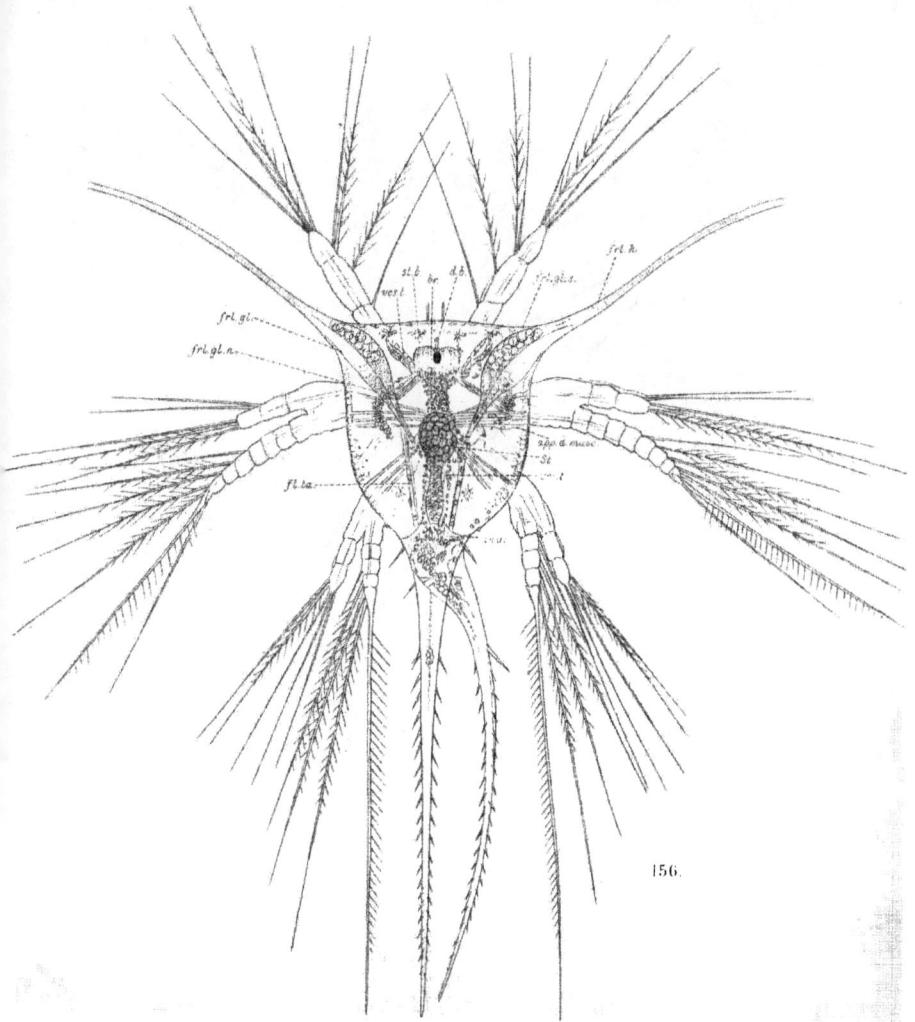

(Fig. 156, Lepas anatifera.)

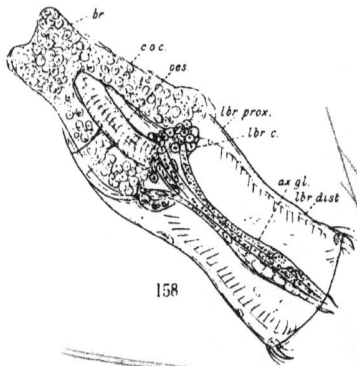

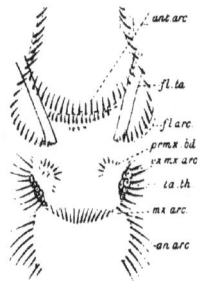

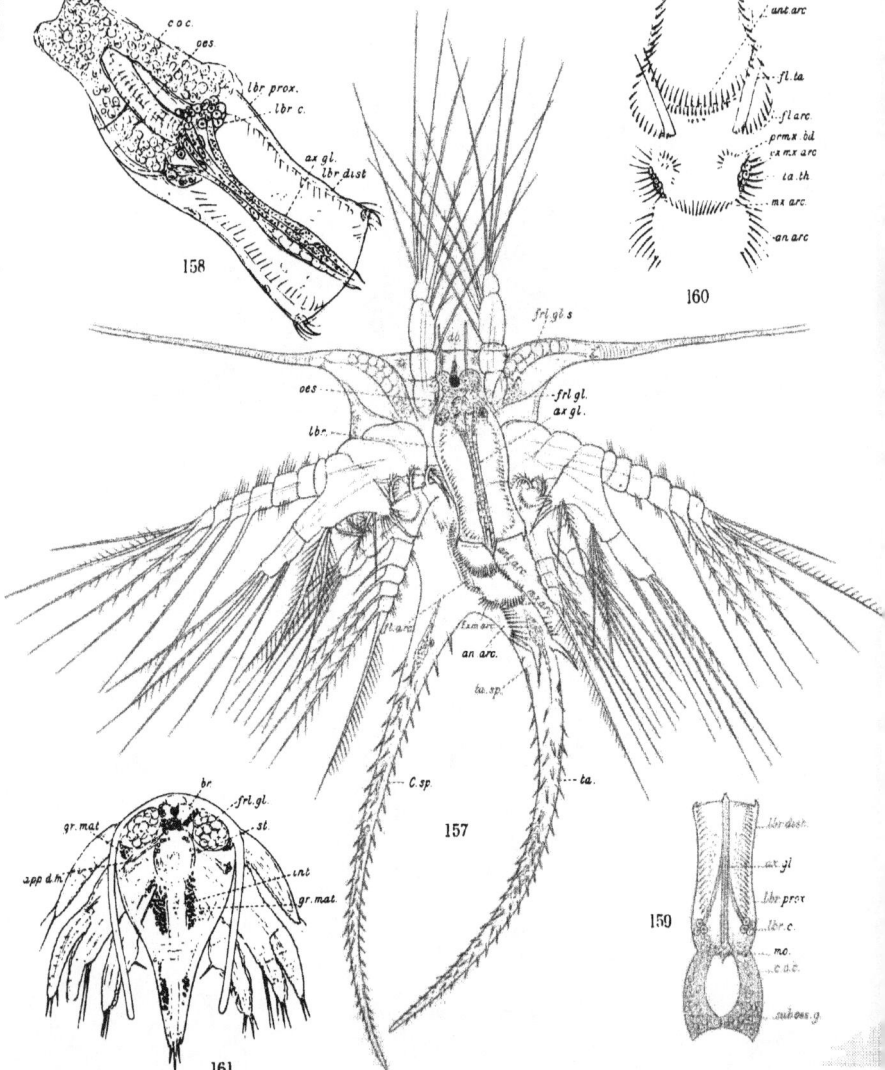

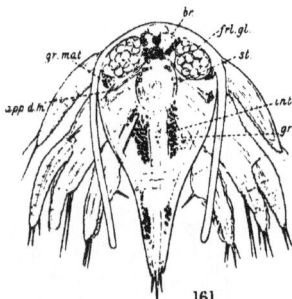

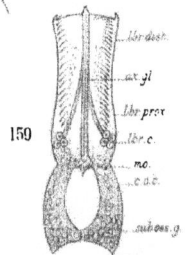

(Fig§ 157–160, *Lepas anatifera* , Fig. 161, *Lepas pectinata.*)

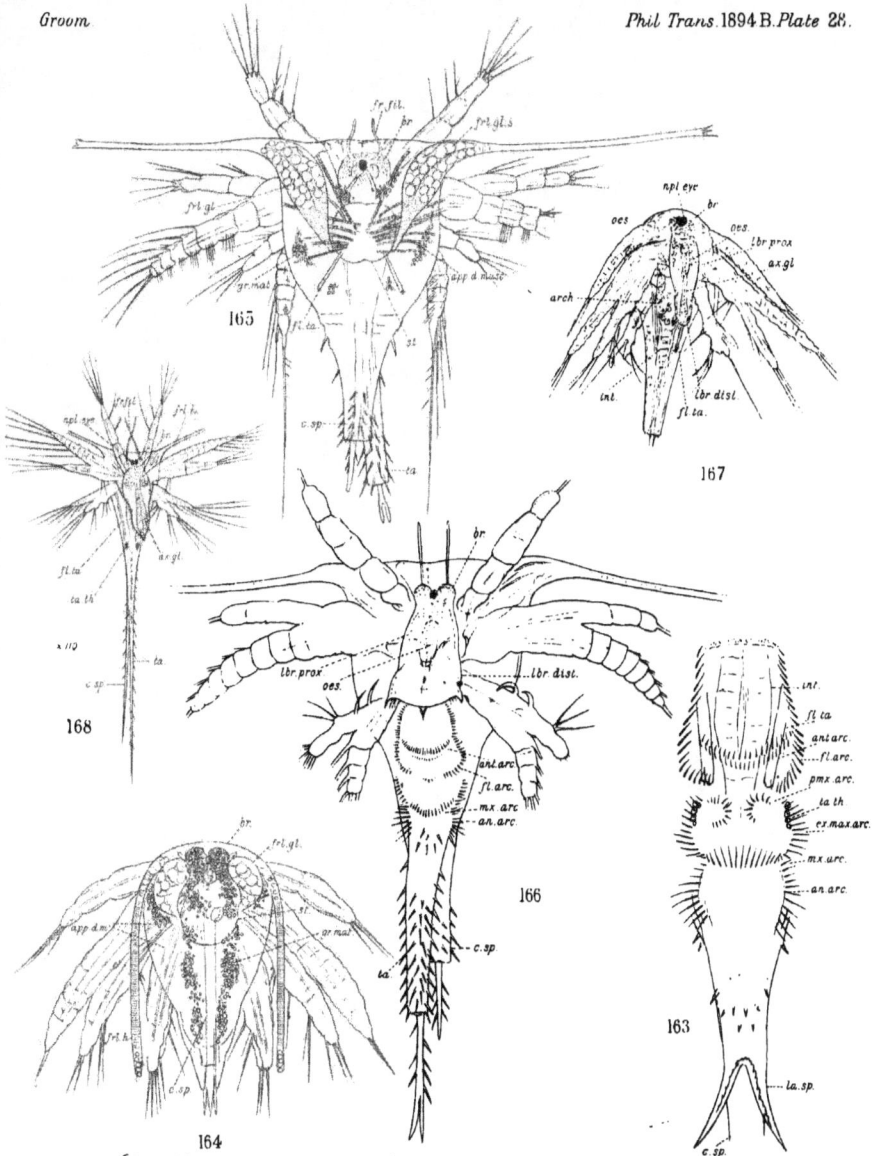

(*Fig. 163, Lepas pectinata,* Fig^s *164–166, Conchoderma virgata.*)

www.ingramcontent.com/pod-product-compliance
Lightning Source LLC
Chambersburg PA
CBHW021933190326
41519CB00009B/1010